Ahmed Snoussi
Mélika Mankai
Hayet Ben Haj Koubaier

Citrus: Biological control of micro-organisms

Ahmed Snoussi
Mélika Mankai
Hayet Ben Haj Koubaier

Citrus: Biological control of micro-organisms

Antimicrobial activity of some essential oils on citrus contamination strains

ScienciaScripts

Imprint

Any brand names and product names mentioned in this book are subject to trademark, brand or patent protection and are trademarks or registered trademarks of their respective holders. The use of brand names, product names, common names, trade names, product descriptions etc. even without a particular marking in this work is in no way to be construed to mean that such names may be regarded as unrestricted in respect of trademark and brand protection legislation and could thus be used by anyone.

Cover image: www.ingimage.com

This book is a translation from the original published under ISBN 978-620-6-69548-6.

Publisher:
Sciencia Scripts
is a trademark of
Dodo Books Indian Ocean Ltd. and OmniScriptum S.R.L publishing group

120 High Road, East Finchley, London, N2 9ED, United Kingdom
Str. Armeneasca 28/1, office 1, Chisinau MD-2012, Republic of Moldova, Europe
Printed at: see last page
ISBN: 978-620-7-30272-7

List of abbreviations

EO: essential oil

DMSO: Dimethyl sulfoxide

GN: nutrient agar

SAB: Sabouraud

E.coli: *Escherichia coli*

B.cereus: *Bacillus cereus*

S.aureus: *Staphylococcus aureus*

S.thyphi: *Salmonella thyphimurium*

S.sonnei: *Shigella sonnei*

P. aeruginosa: *Pseudomonas aeruginosa*

T.spp:

USDA: United States Department of Agriculture

DGPA: Direction Générale de la Production Agricole (General Directorate for Agricultural Production)

INS: Institut national de la statistique

Contents

General introduction

Citrus growing is ranked among the main fruit crops worldwide as well as in Tunisia, this importance is attributed to the major role this crop plays at the economic level; the high nutritional quality of citrus fruits and their richness in vitamins, as well as the favorable pedoclimatic conditions all give one more reason for the development of this crop worldwide (Benaissat F et al.2015) [1].

They represent the world's leading fruit production. The Mediterranean basin ranks second and presents a great diversity of pedoclimatic conditions (Jacquemond et al.2002) [2]

Citrus includes all citrus groups: oranges, clementines, mandarins, lemons and pomelos (Jacquemond et al. 2002) [2].

Infection of citrus fruit with the pathogen can occur both before and after harvest. Citrus fruit infections are initiated by contamination in the field, a few days to several weeks before harvest. These infections are favored by, among other things, a high inoculation rate in the air and high humidity. Deterioration can be limited by unfavorable environmental conditions, and also by resistance of the fruit rind (Douyle et al.1990) [3].

The entry points for pathogens during and after citrus harvesting (storage) are wounds and accidental micro-lesions (FARBER et al.1989) [4].

Food-borne illnesses resulting from the consumption of food contaminated with pathogenic bacteria have been a vital public health concern. Currently, there is a lively debate on the safety aspects of chemical preservatives as they are considered to be responsible for numerous carcinogenic attributes as well as residual toxicity (M.Oussalah et al.2007) [5]

For these reasons, the use of natural products as antibacterial compounds for food preservation is receiving increasing attention due to consumer awareness of natural food products and growing concern about microbial resistance towards conventional preservatives (Sabrine et al.2016) [6]

In fact, these natural products appear to be an interesting way of controlling the presence of pathogenic bacteria and extending the shelf life of processed foods. Among these products, essential oils (EOs) from spices, medicinal plants and herbs have been shown to possess antimicrobial activities and could serve as a source of antimicrobial agents against food pathogens (Sabrine et al.2016) [6].

Essential oils and their compounds are known to be active against a wide variety of microorganisms, including Gram-negative and Gram-positive (Sabrine et al.2016) [6]

Gram-negative bacteria were generally found to be more resistant than Gram-positive bacteria to the antagonistic effects of essential oils due to lipopolysaccharides present in the outer membrane, but this was not always true (Sabrine et al.2016) [6]

The antimicrobial activity of essential oils is attributed to a number of small terpenoids and phenolic compounds (thymol, carvacrol, eugenol), which also in their pure state exhibit strong antibacterial activity (Sabrine et al.2016) [6]

The aim of the project is to meet consumer requirements in terms of food safety and quality, while improving the shelf life of citrus fruits.

The aim of this study is to investigate the antibacterial effect of certain essential oils such as thyme (Thymus vulgaris), rosemary (Romarinus officinalis), black pepper (piper nigrum) and sage (Salvia officinalis), etc., on strains of pathogenic bacteria such as Staphylococcus aureus, Bacillus cereus, Pseudomonas sp and Escherichia coli, which are involved in the degradation of essential citrus fruit quality, as a tool for decontaminating citrus fruit.

This report is divided into three sections:

- The first section is devoted to a bibliographical study of citrus fruits (pathogenic microorganisms as sources of contamination) and essential oils.
- A second section presents the methodology: materials and methods used in this project.
- Finally, a third section is devoted to results and discussion. In this last section, we presented and discussed the antibacterial effect of essential oils on the pathogenic strains that contaminate citrus fruit.

Revue Bibliographique

Chapter 1: CITRUSES

I. General information on citrus fruit

Citrus fruits, also known as hesperidia, are trees producing fruits characterized by a skin surface (zest) rich in essential oil andes, and a pulp organized into quarters including seeds and numerous succulent hairs gorged with juice (Imbert.2005) [7]

The word "citrus", of Italian origin, designates the edible fruits and, by extension, the trees that bear them, belonging to the "Citrus" genus. The main citrus trees cultivated for fruit production are: oranges, mandarins, clementines, lemons and pomelos. All these species belong to the Rutaceae family, which comprises three genera (Poncirus, Fortunelle and Citrus) (Loussert et al.1987) [8].

II. Origin and history

Citrus fruits are native to subtropical Asia, and more specifically to an area stretching from northeast India to northern Indonesia, via Myanmar (Burma) and southern China. Citrus was first imported into the Mediterranean area in the 3rd century B.C., and some authors date it to Alexander the Great's epic voyage to Persia, where the citron tree was cultivated. Called the "Persian Apple" at the time, the citron brought back to Greece quickly conquered the rest of the Mediterranean. It would be several centuries before other citrus varieties were introduced to the West. The primacy of the citron tree in the West and the absence of other citrus fruits over a period extending from Antiquity to the Middle Ages are controversial among historians and archaeologists. Apart from Southeast Asia, the Mediterranean basin is considered the springboard for the spread of citrus cultivation worldwide. Trade with Asia from the 10th century onwards saw the Genoese and Portuguese introduce orange, sour orange and lemon trees to the Mediterranean basin. The Moors introduced orange cultivation throughout the Maghreb and the western Mediterranean (Aissa et al.2021) [9].

III. Productions

1. Global production

According to data from the **US** Department of Agriculture **(USDA)***, world citrus production for all products combined came to over 90 Mt for the 2016/17 season, with a CAGR of 1.2% over the 2007-2017 period. In general, global citrus production can be broken down into four categories as follows:

Table 1: The four categories of citrus fruit

	Part dans la production Mondiale
Oranges	54%
Tangerines, Mandarines	31%
Citrons	8%
Pamplemousses	7%

Source: ONAGRI calculations based on USDA

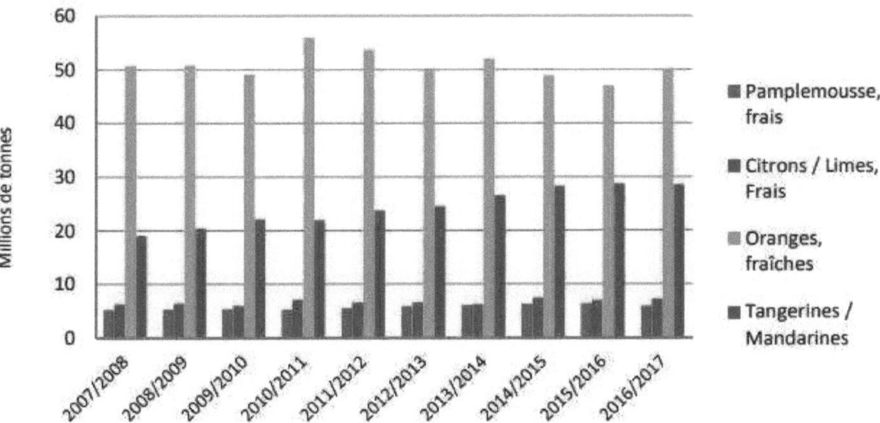

Figure 1*: World production trends by citrus variety (Millions of tons)*

Source : USDA

Over the last decade, tangerine production has increased by 5.2%, from 19 MT in 2007/2008 to 29 MT in 2016/2017. These small citrus fruits are mainly produced in China, Spain, Morocco, Turkey and other Mediterranean countries.

China is the world's leading citrus producer, with a 34% share and a volume of 29.5 Mt, followed by Brazil with a 22% share. The EU ranks 3rd, followed by Mexico (6.7 Mt) and the United States (4.6 Mt). Morocco ranks seventh, followed by Turkey with a share of 1.6%. Tunisia's share of world production is 0.7%.

2. Production in Tunisia

After the record 560 thousand tonnes of the 2016/17 campaign, citrus production for the 2017/2018 campaign is expected to record a significant drop of 38.2% or 346 thousand tonnes due to unfavorable weather conditions that coincided with the flowering period. This drop in production would be due to the decline in Maltese (49%), Clementines (38%) and Navels (31%).

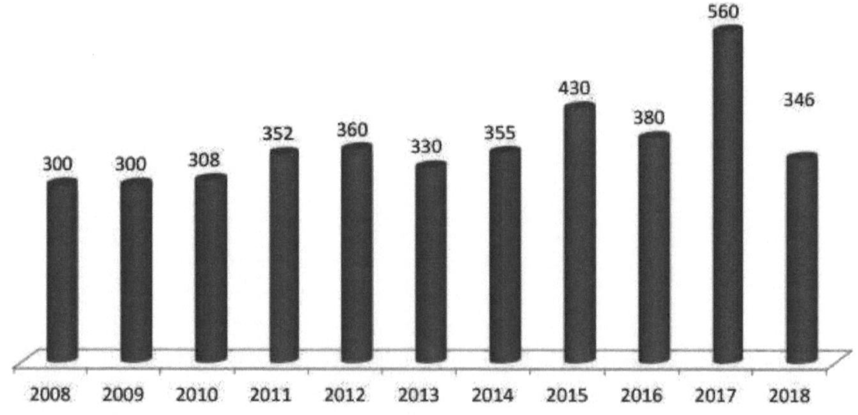

Figure 2: Trend in citrus production in Tunisia (1000 t)

Source: DGPA

In Tunisia, 90% of citrus fruit production is destined for the local market for fresh consumption, which has been growing steadily in recent years.

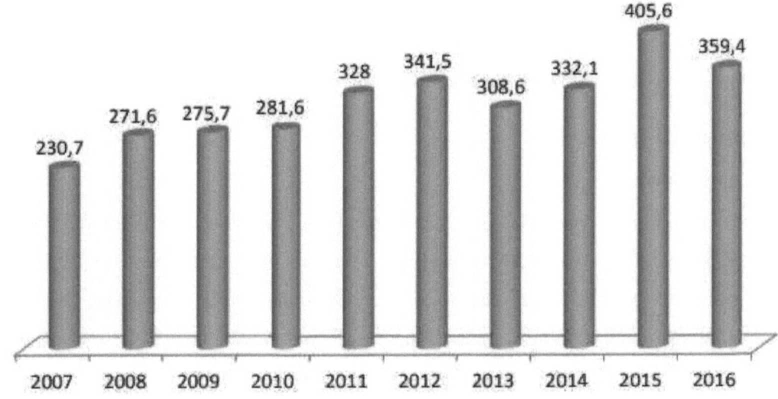

Figure 3:*Citrus consumption trends in Tunisia (1000 tonnes)*

NB: Consumption is equal to the difference between production and exports.

Source: DGPA+INS

IV. <u>Microbiology of orange and orange extracts</u>

Consumption of fruit juices has risen sharply in recent years, as they are rich in minerals and antioxidants essential for good health (Aissa et al.2021) [9].

Fruit juices, like all acidic foods, can be contaminated by acid-tolerant bacteria, yeasts and moulds (Aissa et al.2021) [9].

In fact, fruits and vegetables contain the nutrients required for microbial growth, although the frequency of poisoning from these plants is lower than from other foods (Aissa et al.2021) [9].

Pathogenic bacteria, long shunned in acidic foods such as fruit j u i c e s , have nevertheless been detected there: ***E. coli, salmonella, shigella...*** can survive for days or even weeks in these acidic environments.

1. <u>Alteration flora</u>

1.1. Acetic bacteria

These bacteria are often found on plant surfaces, and can contaminate fruit juices. The most predominant species found on fruit surfaces are Acetobacteraceti and Acetobacterpasteurianus, while yeasts and molds are spoilage flora.

1.2. Pathogenic bacteria

a. Yeasts and molds

Mould contamination of foodstuffs is currently receiving a great deal of attention due to the mycotoxins these micro-organisms are capable of synthesizing. Molds are widely distributed in nature (air, soil, etc.) and can easily contaminate foods during processing. Acidophilic, psychrotrophic or osmophilic yeasts can induce profound alterations in foodstuffs (structure, organoleptic properties, etc.). Most are non-pathogenic (Aissa et al.2021) [9]

b. Escherichia coli

E. coli is the predominant enterobacterium of the digestive tract, and its identification poses no particular problem. in particular E. coli responsible for diarrhoea (Aissa et al.2021) [9].

c. Salmonella sp

Salmonella is a bacterium that contaminates foodstuffs when hygiene rules are not respected, making it a major problem in food microbiology. Salmonellosis remains the most widespread food-borne illness in the world (Aissa et al.2021) [9].

d. Shigella sp

Bacteria of the Shigella genus are strict pathogenic enterobacteriaceae, and the Shigelladysenteriae species is responsible for bacillary dysentery or shegellosis (Aissa et al.2021) [9].

e. Bacillus cereus

Being thermophilic bacteria, their growth temperature range extends from 5°C to 55°C, with an optimum between 30°C and 37°C. Their growth pH is between 4.5 and 9.3.

This ubiquitous species is widespread in nature, found mainly as spores in soil, on plant surfaces, in rivers and in ambient air.

It is also recognized as an undesirable agent in the agri-food industry for its alteration to the organoleptic qualities of foodstuffs. Its ability to sporulate makes it resistant to heat treatment and the formation of biofilms makes it difficult to eradicate (HAOUCHINE et al, 2017) [15].

2. <u>Source of contamination</u>

Pathogenic and spoilage micro-organisms can contaminate fruit and vegetables in a variety of ways, either before or after harvest (Deak et al. 1993) **(19).**

Contamination is thus diverse, affecting the fruit and its environment, production conditions, harvesting, transport, pressing and packaging.

i. Contamination of agricultural crops

The origin of these contaminations can be the plant itself, the soil, fertilizers, animal faeces, contaminated irrigation water, air environment.... Damaged or fallen fruit used in juice production represents a potential risk o f   contamination, since the wounds produced encourage t h e   massive entry and multiplication of microorganisms inside the fruit.

ii. Contamination in the production and storage chain

Any step involved in production is a risk for contamination:

<u>Transporting fruit or squeezed fruit juices</u>: this is also a source of contamination due to the presence of micro-organisms in the transport trucks and their surrounding environment.
fields, where the juices are pressed and concentrated, then reconstituted in another location. This
also minimizes transport costs (AFNOR, 1970) **(21).**

<u>The manufacturing plant or storage area</u>: microorganisms can be introduced into the fruit, from the air, water, the producer himself (hands, hair, etc.), the centenarians used, machines, etc. Hygiene measures must therefore be absolutely rigorous (AFNOR, 1970) **(21).**

Chapter 2: ESSENTIAL OILS

I. General information on medicinal plants

Several definitions have been given to aromatic and medicinal plants (AHP), and the range of these plants is very long and elastic, encompassing most spontaneous plants and many cultivated arboricultural and herbaceous species.

According to (Peyron, 2000), these various plants can be used, in turn or together, for aromatic, medicinal, cosmetic or perfumery purposes. They are used in a variety of forms: unprocessed, processed (dehydrated, frozen), elaborated (extracts, essential oils, oleoresins, isolates). They can also be distinguished according to the organs harvested. To avoid any discrepancies in the understanding of certain key words, we adopt in this chapter the definitions given by the World Health Organization(WHO). (Redouane, 2020)[10].

According to the WHO, "a medicinal plant is a plant which contains, in one or more of its organs, substances which can be used for therapeutic purposes, or which are precursors of hemi-synthetic chemo-pharmaceuticals". This definition makes it possible to distinguish between already known medicinal plants whose therapeutic properties or as a precursor of certain molecules have been scientifically established, and other plants used in traditional medicine. (Redouane, 2020)[10]

II. Essential oils

1. Definition

Essential oils are products of generally fairly complex composition, containing volatile principles found in plants and more or less modified during preparation (Redouane, 2020) [10].

More recently, AFNOR standard NF T 75-006 (October 1987) defined an essential oil as follows: Product obtained from a vegetable raw material, either by steam distillation, or by mechanical processes from the epicarp of citrus fruits, or by dry distillation.

Essential oil (EO) is the liquid obtained from a plant by distillation or chemical extraction using solvents. Despite its name, it is not always a greasy or oily liquid. Widely used in alternative medicine (aromatherapy), essential oils are full of virtues, and their health benefits are well established. It is estimated that around 10% of plants can produce essential oils (Nadjib et *al.* 2019) [20].

2. Location and performance

A priori, all plants have the ability to produce volatile compounds, but usually only in trace amounts. Among plant species, only 10% are said to be "aromatic" (Nadjib et al. 2019) [20].

The ability to accumulate EO is, however, the property of certain plant families distributed throughout the plant kingdom, as well represented by the gymnosperm classes Cupressaceae (cedar wood) and Pinacea (pine and fir) as that of the angiosperms (Nadjib et al.,2019)[20].

EOs are natural secretions elaborated by the plant and contained in cells or parts of the plant such as those of flowers (rose), flowering tops (lavender), leaves (lemongrass), barks (cinnamon), roots (iris), fruits (vanilla), bulbs (garlic), rhizomes (ginger) or seeds (nutmeg) (Nadjib et al.,2019)[20] .

Essential oils are produced in specialized glandular cells covered by a cuticle (Redouane, 2020) [10].

They are then stored in essential oil cells (Lauraceae or Zingiberaceae), secretory hairs (Lamiaceae), secretory pockets (Myrtaceae or Rutaceae) or secretory ducts (Apiacieae or Asteraceae) (Redouane, 2020) [10].

They can also be transported into the intracellular space when essence pockets are located in internal tissues (Redouane, 2020) [10].

On the storage site, essential oil droplets are surrounded by special membranes made of highly polymerized hydroxylated fatty acid esters, combined with peroxide groups (Redouane, 2020) [10].

Due to their lipophilic nature and therefore extremely low permeability to gases, these membranes greatly limit the evaporation of essential oils and their oxidation in air (Redouane, 2020) [10].

3. Chemical composition

The chemical composition of aromatic plants is complex. The number of chemically different molecules that make up an essential oil is variable, with a great diversity of compounds (up to 500 different molecules in Rose essential oil) (Redouane, 2020) [10]

Alongside the majority compounds (generally between 2 and 6), there are minority compounds and a number of trace constituents (Redouane, 2020) [10].

Relatively low molecular weight (terpenes: 136 a.m.u., terpinols: 154 a.m.u., and sesquiterpenes: 200 a.m.u.), which gives them a volatile character and is the basis of their olfactory properties. HE is made up of two fractions (Redouane, 2020) [10].

The first fraction, known as the volatile fraction (VOC), is present in different plant organs depending on the family; this fraction is made up of secondary metabolites that constitute the essential oil (Redouane, 2020) [10].

The plant's second non-volatile fraction, non-volatile organic compounds (NVOCs), is essentially made up of coumarins, flavonoids, acetylenic compounds and polyphenols that play a fundamental role in the plant's biological activity (Redouane, 2020) [10].

Aromatic plants are unique in that their secretory organs contain cells that generate secondary metabolites, clearly showing how highly volatile molecules are synthesized from methyl-2-buta-1,3-diene (isoprene) units, and the addition reactions of these units lead to terpenes, sesquiterpenes, diterpenes and their oxidation products such as alcohols, aldehydes, ketones, ethers and terpene esters (Redouane, 2020) [10].

All these products are accumulated in secretory cells, giving the plant a characteristic odor (Redouane, 2020) [10].

3.1. Terpenoids
Terpenes are highly volatile molecules common in nature, especially in plants, where they are the main constituents of essential oils (Redouane, 2020) [10].

Terpenes are derived from the coupling o f at least 2 5-carbon isoprenic subunits (Redouane, 2020) [10].

3.2. Aromatic compounds

Another class of volatile compounds frequently encountered are aromatic compounds derived from phenylpropane (Redouane, 2020) [10].

This class includes well-known odorant compounds such as vanillin, eugenol, anethole, estragole and many others (Redouane, 2020) [10].

They are more frequent in Apiaceae essential oils (parsley, anise, fennel, etc.) and are characteristic of those of clove, vanilla, cinnamon, basil, tarragon, etc. (Redouane, 2020) [10].

4. <u>Physico-chemical properties of essential oils</u>

Essential oils are liquids at ordinary temperatures, with a very pronounced aromatic odor, generally colorless or pale yellow with the exception of a few essential oils such as Yarrow oil and Matricaria oil (Redouane, 2020) [10].

These are characterized by a blue to greenish-blue coloration, due to the presence of azulene and chamazulene (Redouane, 2020) [10].

Most essential oils have a density lower than that of water and are steam-entrainable; there are, however, exceptions such as Sassafras, Clove and Cinnamon essential oils whose density is higher than that of water (Redouane, 2020) [10].

5. <u>Biological activities</u>

Essential oils are known for their antiseptic and antimicrobial properties, and many of them have antitoxic, antivenom, antiviral, antioxidant and antiparasitic properties. More recently, they have also been recognized for t h e i r anticancer properties (Lahlou et al. 2004).

5.1. Antifungal power

In the phytosanitary and agri-food sectors, essential oils or their active compounds could also be used as protective agents against phytopathogenic fungi and microorganisms invading foodstuffs (LisBalchin, 2002)[22].

In the phytosanitary and agri-food sectors, essential oils or their active compounds could also be used as protective agents against phytopathogenic fungi and microorganisms invading foodstuffs (Abed et *al.* 2021) [23].

5.2. Antibacterial power

According to (Benayad, 2008), phenols (carvacrol, thymol) have the highest antibacterial coefficient, followed by monoterpenols (geraniol, menthol, terpineol), aldehydes (neral, geranial), etc. Example: oregano EO, thymol thyme EO (doctissimo.fr, 04-2020).

The HEs (essential oils) most studied for their antibacterial properties belong to the Labiatae: oregano, thyme, sage, rosemary and clove are aromatic plants with essential oils rich in phenolic compounds such as eugenol, thymol and carvacrol (Abed et *al.* 2021) [23]. These compounds have strong antibacterial activity. Carvacrol is the most active of all (Abed et *al.* 2021) (23).

The compounds with the greatest antibacterial efficacy and broadest spectrum are phenols: thymol, carvacrol and eugenol. Carvacrol is the most active of all, and is recognized for its non-toxicity; it is used as a preservative and food flavoring in beverages, sweets and other preparations. Thymol is the active ingredient in mouthwashes. Eugenol is used in cosmetics, food and dental products. These compounds have an antibacterial effect against a wide spectrum of bacteria, including Escherichia coli, Bacillus cereus, Listeria monocytogenes, Salmonella enterica, Clostridium jejunii, Lactobacilluss sakei, Staphylococcus aureus and Helicobacter pylori (Fabian et al., 2006) (24).

6. **Essential oil extraction methods**

=> Extraction by hydrodistillation.

=> Steam extraction.
=> Hydro-diffusion.

=> Cold expression.

=> Solvent extraction.
=>Extraction by fats.

=>Microwave extraction.

A hydrodistillation method was used.

6.1. Extraction by hydrodistillation

The plant material i s immersed directly in a water-filled alembic placed over a heat source, then brought to the boil. The heterogeneous vapors are condensed in a condenser, and the essential oil is separated from the hydrosol by a simple difference in density, the essential oil being lighter (Paupardin et al., 1990) (25).

7. Toxicity of essential oils

Essential oils are high-risk products. Toxicity stems from the presence of certain aromatic molecules for which risks have been identified through testing: the ketone family: (neurotoxicity and abortifacient risk), the phenol and aldehyde family : (dermocausticity, hepatotoxicity, irritation of respiratory mucosa, triggering asthma attacks), the furocoumarin and pyrocoumarin family: (erythematous reactions under prolonged sunlight), the monoterpene family: (inflames and deteriorates, nephrons) (Aomari et al.,2018) **(24)**.

In today's world of natural products, these substances should not be abused. As with a medicine, for each essential oil there is a balance between benefit and risk that must also be considered according to the subject (Aomari et al. 2018) **(24)**.

8. <u>Regulations</u>

As soon as the manufacturer indicates that an essential oil is intended for ingestion, it must be of food-grade quality. There are two types of use in this field:

➤ Oils marketed for aromatic use

Many essential oils can be used in cooking. European regulations governing flavors contain a number of provisions, including those relating to labeling and the obligations of those responsible for first marketing (Regulation (EC) No 1334/2008).

Essential oils from plants whose use is recognized for the manufacture of flavorings may therefore be used in food, provided that their dose of use is compatible with their use as flavorings (of the order of 2% maximum) (Regulation (EC) No 1334/2008).

➤ Oils marketed as dietary supplements

Certain essential oils can be used to supplement a normal diet. In this case, regulations require them to be declared to the DGCCRF. Health claims on these products are also subject to prior authorization (Regulation (EC) no. 1829/03). Depending on its composition and presentation, an EO intended for the consumer may be considered a medicine, a cosmetic or a foodstuff.

Materials & Methods

I. <u>Biological materials</u>

The fruits, the subjects of our study, come from fruit plants belonging to the following species:

* **Clementine** (Citrus clementina Hort. ex. Tan.), **Mandarin** (Citrus reticulata Blanco), **Orange** (Citrus sinensis Osb), **Lemon** (Citrus limon L.), **Pomelo** (Citrus paradisi Macf.) (Khaled.2020) [16]

1. <u>Sampling</u>

Citrus fruit: for citrus control we have chosen the following levels:

- citrus receiving station

- brushing station

-wash before and after exit.

-citrus storage (moldy citrus)

1 kg of fruit is taken as a representative sample. The fruit is washed with sterile distilled water at a rate of 250 ml per 1 kg sample.

Analysis of the wash water gives us an indication of the degree of contamination of the fruit by pathogenic strains (bacteria or fungi). (Khaled.2020) [16]

2. <u>Microbiological analysis of citrus fruit</u>

2.1. **Enumeration of juice and fruit microflora.**

A series of dilutions of semi-finished juice or wash water is performed to obtain a countable bacterial suspension on solid medium in Petri dishes. Dilutions of
10 in 10 are carried out. Inoculation can also be carried out using the mass inoculation method, but this method underestimates population size, as oxygen-demanding strains will not grow. In principle, each living strain introduced into the mass of a favorable agar medium gives rise to a colony that can be seen with the naked eye. Consequently, if a product or its dilution is inoculated into this culture medium, the number of colonies developed corresponds to the number of micro-organisms present in the inoculated volume (Khaled.2020) [16].

i. Yeast and mold testing

❖ How it works

Enumeration of fungal flora was carried out on Sabouraud. A 10 ml sample of

After solidification, these dishes are inoculated with 0.1 ml of the surface juice solution and incubated at 37°C for 3-10-15 days.

ii. Testing for Escherichia coli

❖ How it works

Pour 15 ml of Hekteon agar into empty petri dishes for liquefied enumeration at 37°C, mix thoroughly and allow to solidify.

After solidification, these dishes are inoculated with 1 ml of the surface juice solution (Aissa et al.2021) (9).

iii. Testing for Staphylococcus aureus

❖ How it works

Pour 15 ml of Baird Parker agar into petri dishes for liquefied enumeration at 37°C, mix thoroughly and allow to solidify.

After solidification, these dishes are inoculated with 0.1 ml of the juice solution on the surface (Aissa et al.2021) (9)

iv. Testing for Salmonella sp

❖ How it works

Salmonella enrichment is performed in selenite enrichment broth. On the other hand, SS agar medium is melted and poured into petri dishes (two dishes per dilution), followed by surface seeding with Hekteon medium. Incubate inoculated Petri dishes at 37°C for 24 to 48 hours (Aissa et al.2021) (9)

v. Testing for Shigella sp

❖ How it works

Hekteon agar was initially used for both Salmonella and Shigella, but Shigella grow much better on Hekteon agar. (Aissa et al.2021)(9).

2.2. Identification

Bacteria are identified in four stages:

- Examination of the macroscopic characteristics of the bacterial colony (shape, relief, odour, contour, size and color);
- Microscopic examination (Gram stain, germ shape);
- Research into biochemical characteristics (Catalase, Oxidase, etc.);
- Confirmation of certain strains by api gallery.

2.2.1. Macro and microscopic study

Colony observation

After incubation, all macroscopic characteristics of the colonies are noted.

- Colony shape: circular, irregular or rhizoid.
- Appearance: punctiform (< 1mm diameter), medium, large or invasive.
- Opacity: transparency, translucency.
- Elevation: flat, convex, centered, raised colony.
- Surface: smooth, rough, blunt, shiny, dry, powdery, wrinkled, creamy.
- Edges: whole, wavy, lobed, toothed, rhizoid, crenellated.
- Consistency: viscous or granular.
- Odor: presence or absence of characteristic odor.

Each colony type is sterile-potted into a tube of sterile nutrient broth, after purification.

Microscopic observation

Once a bacterium has been isolated, it needs to be identified. An initial classification is made on the basis of morphological and cultural characteristics, then species identification is carried out using identification media. Some identification media are specific to certain germs, while others can be used for many germs. The first condition is to have a germ in its "pure" state, and an abundant and sufficient culture to inoculate different identification media.

2.2.2. Gram staining

Bacteria unbleached by alcohol are said to be Gram-positive; they appear purple, while Gram-negative bacteria, bleached by alcohol and recolored by contrast dye, appear pink. These two behaviors result from a fundamental difference in wall composition and structure. Unlike the walls of Gram-negative bacteria, those of Gram-positive bacteria form a barrier preventing alcohol from staining the cytoplasm where the reaction takes place.

Smear preparation

- Place 1 drop of sterile water on a glass slide using a sterile pipette

- Remove a bacterial or fungal colony using a sterile loop, and mix it with the drop of water.

- Allow to air dry.

➤
Perform a Gram stain

- The fixed smear is stained for 1 minute with a gentian violet solution (60s).

- It is then rinsed under a trickle of clear water.

- An iodine-iodide solution of Lugol's is added to act as a "mordant", and the smear is kept in this medium for 1 minute.

- After washing with clean water

- an alcohol-acetone mixture is poured drop by drop onto the inclined slide (for 20 seconds).

- As soon as the solvent runs clear, stop its action immediately by rinsing thoroughly with water and draining well.

- The smear is then contrast-stained by treating it with a safranin solution for 30 seconds,

- rinsed thoroughly with clean water

- and air-dried, or gently dried between two sheets of blotting paper.

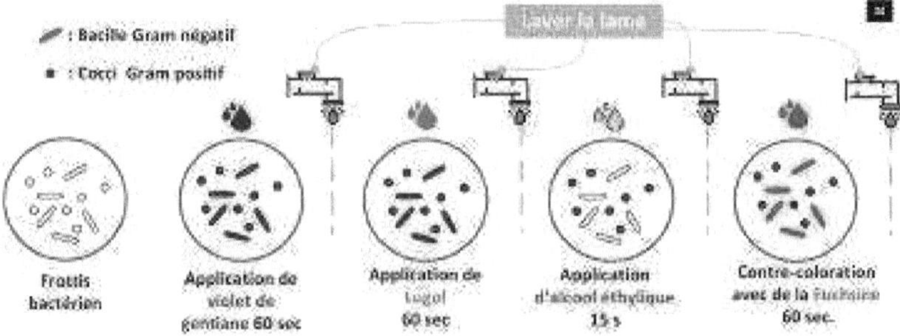

Figure 4: gram staining protocol

➢
Observe the stained smear under the microscope (40x then 100x).

Bacteria unbleached by alcohol are said to be Gram-positive; they appear purple, while Gram-negative bacteria, bleached by alcohol and recolored by contrast dye, appear pink. Yeasts and mycelial filaments appear Gram-positive.

2.2.3. Biochemical activities

Enzyme activity tests to help guide diagnosis. Perform the appropriate test (catalase or cytochrome oxidase) depending on the type of bacteria you have in culture: gram-positive shell or gram-negative bacillus.

Catalase production

Principle

- Highlighted by oxygen production (positive test) when bacillus or Gram+ cocci bacteria are present.

Usefulness of the test

- differentiation of Gram-positive shell bacteria

How it works

- the test consists in transferring a colony fragment to a drop of hydrogen peroxide placed on a glass slide: the presence of catalase gives rise to oxygen bubbles.

$2H_2O_2 \blacktriangleright 2H_2O + O_2$

Cytochrome oxidase production

Principle

This test establishes whether a bacterium contains a certain type of cytochrome in its respiratory chain. Oxidation of chromogenic substrates such as tetra methyl-p-phenylenediamine produces an intense violet color.

Usefulness of the test

Phenotypic identification according to enzyme production: differentiation of Gram- bacilli

How it works

A paper strip impregnated with the reagent, N dimethyl paraphenylene diamine, is used, on which a colony fragment is spread with an oese. Species containing the oxidase give a positive, violet-colored reaction within 30 seconds.

⬥ Identification of Staphylococci: coagulase

This test, which highlights the ability of bacteria to coagulate plasma, is the main test characterizing S.aureus. It therefore differentiates Staphylococcus aureus from coagulase-negative Staphylococcus. The detection test consists of incubating a mixture of rabbit plasma and the strain to be tested for 4 hours at 37°C. The result is a clot. The appearance of a clot is observed by tilting the tube at 90°C.

⬥ Yeast identification using character galleries

Yeasts can be identified on the basis of biochemical characteristics linked to the production of species-specific enzymes. The API system consists of an identification gallery featuring a number of standardized, miniaturized biochemical tests.

In our study we will identify the yeasts isolated from citrus fruits by the api Candida gallery.

❖ How it works

■ Colony selection

- Check microscopically that the strain under study is a yeast. The following media can be used to isolate colonies before using the

API Candida gallery: Sabouraud agar

■ Preparing the gallery

- Put the bottom and lid of an incubation dish together and distribute about 5 ml of water [demineralized, distilled, or any water without additives or chemical substances likely to release gases (e.g. Cl2, CO2, etc.)] in the cells to create a humid atmosphere.

- Place the gallery in the incubation box.

■ Inoculum preparation

- Open an ampoule of API NaCl 0.85% Medium (2 ml)

- Using a pipette or swab, pick one or more identical, well-isolated colonies and make a suspension with an opacity equal to that of the McFarland 3 standard: evaluate by comparison with an opacity control or densitometer. Preferably use young cultures (18-24 hours).

- Homogenize the yeast suspension well. This suspension must be used extemporaneously.

■ Gallery inoculation

- Dispense the previous yeast suspension into the tubes only, avoiding bubbles (to do this, tilt the incubation dish forwards and place the pipette on the side of the dish).

- Cover the first 5 tests (GLU to RAF) and the last test (URE) with kerosene oil (tests underlined) immediately after gallery inoculation. NOTE: The quality of the filling is very important: tubes that are under- or overfilled will give false positive or negative results.
- Close the incubation box.
- Incubate 18-24 hours at 36°C ± 2°C in an aerobic atmosphere.

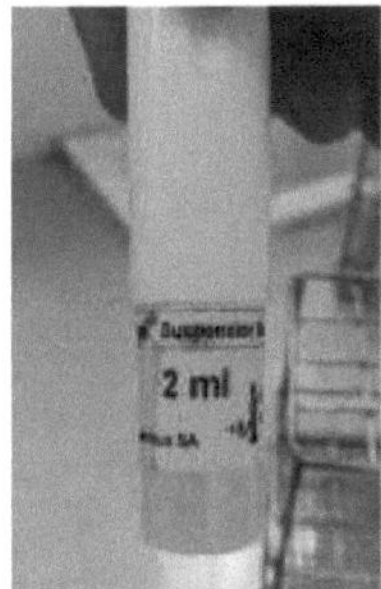 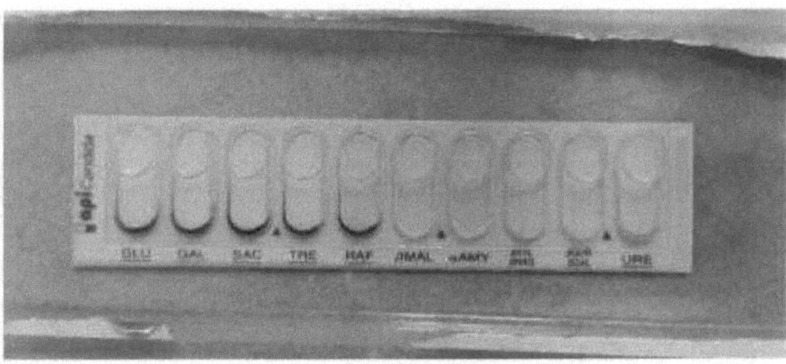

Figure 5: identification of yeasts by api Candida gallery

3. Qualitative assessment of the antimicrobial activity of essential oils: Aromatogramme

3.1. Disc distribution method

The agar disk diffusion test is the official method used in many clinical microbiology laboratories for routine antimicrobial susceptibility testing. In this procedure, agar plates are inoculated with colonies of the test microorganism. Next, filter paper discs (approx. 6 mm in diameter) impregnated with HE at the desired concentration, are placed on the agar surface. Petri dishes are incubated at 37°C for 18 to 24 hours. EO diffuses into the agar and inhibits germination and growth of the test microorganism, then the diameters of the inhibition growth zones are measured. This clear zone around the discs is proportional to the antibacterial activity of the essential oil. It provides qualitative results, classifying bacteria as sensitive, intermediate or resistant. (Alexandra.2020) (17).

3.2. Agar well diffusion method

Similar to the procedure used in the disk diffusion method, the surface of the agar plate is inoculated by spreading a volume of the microbial inoculum over the entire agar surface.

Next, a hole with a diameter of around 6 mm is aseptically perforated and a volume (20-100 µL) of antimicrobial agent or extract solution at the desired concentration is introduced into the well. The agar plates are then incubated under conditions appropriate to the micro-organism being tested. The antimicrobial agent diffuses into the agar medium and inhibits the growth of the microbial strain tested (Alexandra.2020) [17]

a. Microbial strains

The following germs have been tested for the antimicrobial activity of essential oils:

Staphylococcus aureus Pseudomonas aeruginosa Escherichia coli Salmonella thyphimurium
Shigella sonnei
Bacillus cereus
Yeast (to be identified)

b. Re-isolation of microbial strains

In order to obtain pure, young microbial strains, regular re-isolations were carried out using the streak method, with the pasteur pipette on Hekteon medium for Pseudomonas aeruginosa and Escherichia coli, Chapman medium for Staphylococcus aureus and Sabouraud medium for yeasts and molds. (Hessas et al .2020) [11].

c. Preparing EO dilutions

Essential oils used :
- *Thyme EO (Thymus vulgaris)*
- *Sage EO (Salvia officinalis)*
- *Black pepper EO (piper nigrum)*
- *Myrtle EO (Myrtus communis L)*
- *Rosemary EO (Rosmarinus officinalis)*
- *Caraway EO (Carum carvi L)*
- *Garlic EO (Allium sativum)*
- *Clove EO (Syzygium aromaticum)*

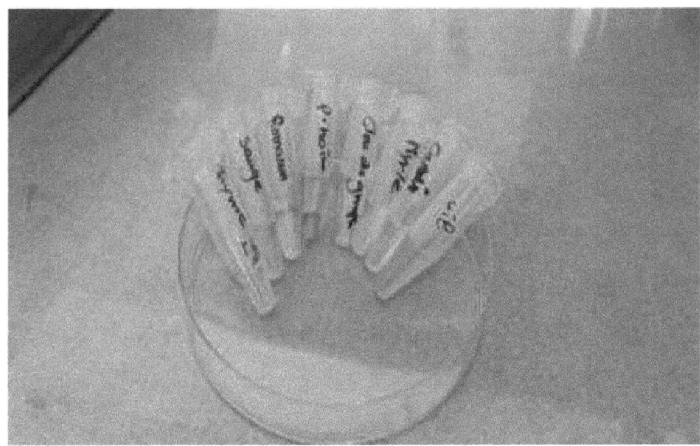

Figure 6: Essential oils

Preparing EO dilutions

A series of dilutions of the 8 essential oils in DMSO (dimethyl sulfoxide) was carried out, starting with a **1/2** dilution up to a 1/8 dilution in sterile glass tubes:

- The first contains 500 µl of essential oil and 500 µl of DMSO.

- 500 µl of the first dilution is transferred to the second tube **(1/4)**, to which 500 µl of DMSO is added, then shake.

- 1/8, 1/16, 1/32 dilutions, are prepared in the same way as shown in **figure 7.**

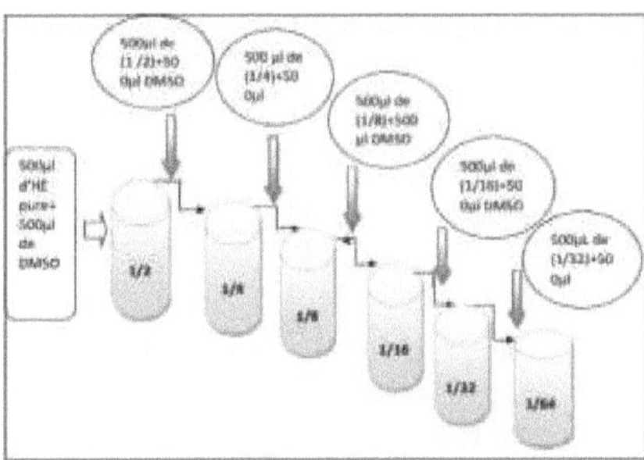

Figure 7: *Preparation of different dilutions*

d. Preparation of microbial suspensions

From a pure culture of the bacteria to be tested on isolation medium, scrape off a few well-isolated, perfectly identical colonies using a sealed pasteur pipette.

- Unload the pasteur pipette into 5 ml sterile physiological water.

- homogenize the bacterial suspension

-

The cell density of a bacterial suspension is adjusted in the course of a
simple comparison with the turbidity of the standard. This can be done by direct visual
comparison, or by measurement in a spectrophotometer.

-

Using a liquid bacterial culture aged 18 to 24 h, we adjusted

the opacity of the suspension to obtain an optical density of between 0.08 and 0.1 at a

wavelength of 600 nm (approx. 10 CFU/ml).

e. Seeding

The culture medium used is nutrient agar (NA), Sabouraud (SAB) for yeasts and molds, which are the most

commonly used media for sensitivity testing to antibacterial agents.

- Dip a sterile swab into the bacterial suspension (to avoid contamination of the operator and bench).

- Wring it out by squeezing it firmly, turning on the inside wall of the tube, to unload it as much as
possible.

- Rub the swab over the entire dry agar surface, from top to bottom, in tight ridges.

- Repeat the operation three times, turning the Petri dish 60° each time, not forgetting to rotate the
swab on itself.

- Finish inoculation by swabbing the periphery of the agar.

\-   When inoculating several Petri dishes, the swab must be reloaded each time.

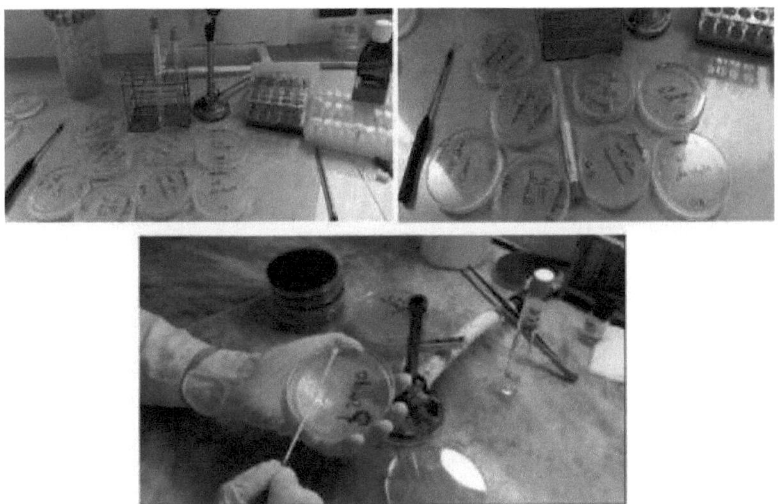

Figure 8: Inoculation of bacterial suspensions

f. Well perforation

A hole approximately 6 mm in diameter is aseptically perforated and a volume (10-20 µL) of antimicrobial agent or extract solution at the desired concentration is introduced into the well.

g. Negative control

In each petri dish, a well was filled with DMSO.

h. Incubation

The plates are left to diffuse for 2 hours at 4°C, incubated in an oven at 37°C for 24 h for the bacterial strains, and for 48 to 72 h for the fungal strain. Three trials are carried out for each test.

i. Expression of results

The absence of microbial growth results in a halo around the wells, whose diameter was measured with a caliper (including the 6 mm well diameter).

A scale for estimating the antimicrobial activity of an essential oil, based on the diameters of the zones of inhibition (D), distinguishes 5 classes of EO.

- Very strong inhibitor: $D \geq 30$ mm

- Highly inhibiting: 21 mm $\leq D \leq 29$ mm

- Moderately inhibiting: 16 mm $\leq D \leq 20$ mm

- Slightly inhibiting: 11 mm $\leq D \leq 16$ mm

- Non-inhibiting: $D \leq 10$ mm

The sensitivity of strains to antimicrobial agents was classified according to the inhibition diameters of the zones of inhibition according to (Djeddi et Al, 2007) [26] as follows:

- (-) resistant strain (D<8 mm)
- (+) sensitive strain (9mm $\leq D \leq 14$mm)
- (+ +) very sensitive strain (15mm $\leq D \leq 19$ mm)
- (+ + +) extremely sensitive (D >20 mm)

4. Quantitative assessment (CMI)

The Minimum Inhibitory Concentration (MIC) corresponds to the lowest concentration of the essential oil that inhibits all microbial growth visible to the naked eye. It is calculated for each species that showed sensitivity to the essential oil in the previous test, using the solid dilution technique (BOUALEM et al.2016) [27].

Preparation of microbial suspensions

They have been prepared as before

Preparing essential oil dilutions

- A series of dilutions of thyme EO in the agar medium is carried out, starting with a 2% dilution up to a 0.03% dilution. They are prepared as follows:

- Incorporate 1 ml of HE into 49 ml of liquefied GN (or Sabouraud) agar medium (in a water bath (95°±2°C) then cooled to 40°±2°C).

- Shake to homogenize the mixture. A 2% volume-to-volume dilution (D0) (v/v) was then obtained.

- Take 25 ml of the previous mixture (D0) and divide into two 12.5 ml petri dishes.

- Add 25 ml of GN (or Sabouraud) liquefied agar medium to the remaining 25 ml of 2% D0 to obtain a 1% dilution (D1).

- Take 25 ml of D1 and divide between two 12.5 ml petri dishes, then add 25 ml of GN (or Sabouraud) liquefied agar medium to the remaining 25 ml of D1 to obtain a 0.5% dilution (D2).

- - Follow the same steps to make the following dilutions: D3 at 0.25%, D4 at 0.125%, D5 at 0.06% and D6 at 0.03%.

- also prepare two boxes without essential oils as negative controls, one with 12.5ml GN and the other with 12.5ml Sabouraud.

Figure 9: Preparation of HE dilutions (MIC)

⬇ Spreading microbial suspensions

After the media had solidified, the various microbial suspensions were spread out using a swab on the GN and a rake on the SAB.

⬇ Incubation

GN plates were incubated at 37°C for 24 hours and SAB plates for 48-72 hours.

⬇ Reading

The MIC is the lowest concentration of essential oil that inhibits bacterial or fungal growth visible to the naked eye.

5. <u>Statistical analysis</u>

The inhibition diameter and MIC values of all samples studied were compared by one-way analysis of variance (ANOVA) using SPSS statistics 20 software. Significant differences ($P < 0.05$) between essential oils were detected using Duncan's multiple range tests.

Results and discussion

1. Results

1.1. Microbiological analysis

1.1.1.   Bacteria

The flora present in the juice of different stations is nothing other than the result of a variety of micro-organisms originating from the fruit itself, the storage atmosphere, the processing and packaging chain, the water and the personnel. The microbiological quality of the product intended for consumption is a vital component in guaranteeing hygienic safety.

The strains isolated and their macro-microscopic and biochemical characteristics are shown in **Table 2** and **Appendix 1.**

Table 2: Strain characteristics

Strains	Shape	Gram	Catalase	Oxidase	Coagulase
S.aureus	Coccus	Gram $^+$	Catalase +	OX	coagulase+
E. coli	Bacillus	Gram $^-$	Catalase +	OX	
Salmonella thyphimurium	Bacillus	Gram-	Catalase +	OX	
Shigella sonnei	Bacillus	Gram-	Catalase +	OX	
Pseudomonas aeruginosa	Bacillus	Gram $^-$	Catalase +	OX-	
Bacillus cereus	Bacillus	Gram++ (grams)	Catalase +	OX-	

1.1.2.   Yeast

Reading the gallery

According to **appendix 2,** the yeast identified is **"*Trichosporon spp*".**

Trichosporon spp is a yeast-like fungus belonging to the Trichosporonaceae. The pathogen is able to adhere to and form biofilms.
Trichosporon spp. are ubiquitous and can be found in soil, plants and water.
The pathogen can also be found in different species of animals, such as bats, birds, pets and livestock.

Trichosporon spp can cause superficial infections such as piedra blanche (fungal infection of the hair shaft) or onychomycosis (fungal infection of the nails). The pathogen also causes an invasive infection known as trichosporonosis. Trichosporonosis occurs mainly in people with weakened immune systems.

1.2. Qualitative assessment of the antibacterial activity of essential oils: Aromatogramme

The agar well diffusion method enabled us to demonstrate the antimicrobial power of EO on microorganisms. Strain sensitivity is reflected in the appearance of an inhibition halo around the wells, with inhibition zones ranging from 4 to 48 mm, indicating that the strains tested are not all equally sensitive to EO. Values shown are averages of three measurements.

Essential oil concentration is related to zones of inhibition. The higher the concentration, the greater the zone of inhibition.

The results of the EO aromatogram are shown in **Table 3.**

Table 3: Inhibition zone diameters

strains	Diameter of inhibition zones (mm) [a]								
	Thymus vulgaris			*Salvia officinalis*			*Allium sativum*		
	Dilutions			*Dilutions*			*Dilutions*		
	1/2	1/4	1/8	1/2	1/4	1/8	1/2	1/4	1/8
S.aureus	40±1	38±1	32±1	20±1	14±1	13±1	18±1	15±1	6±1
E. coli	32±1	28±1	22±1	10±1	9±1	7±1	12±1	10±1	9±1
Salmonella thyphimurium	25±1	18±1	14±1	13±1	9±1	6±1	22±1	19±1	15±1
Shigella sp	32±1	30±1	25±1	15±1	12±1	7±1	16±1	16±1	12±1
P. aeruginosa	8.5±1	7±1	7±1	10±1	6±1	5±1	10±1	9±1	9±1

| B.cereus | 26±1 | 22±1 | 18±1 | 16±1 | 14±1 | 13±1 | 23±1 | 19±1 | 13±1 |
| richosporon spp | 48±1 | 43±1 | 43±1 | 20±1 | 18±1 | 14±1 | 36±1 | 35±1 | 33±1 |

strains	Diameter of inhibition zones (mm) [a]								
	Myrtus communis L Ro			smarinus officinalis			piper nigrum		
	Dilutions			Dilutions			Dilutions		
	1/2	1/4	1/8	1/2	1/4	1/8	1/2	1/4	1/8
S.aureus	25±1	22±1	20±1	15±1	16±1	13±1	15±1	11±1	6±1
E. coli	13±1	11±1	9±1	9±1	11±1	10±1	10±1	13±1	16±1
Salmonella thyphimurium	12±1	10±1	9±1	18±1	14±1	8±1	15±1	13±1	11±1
Shigella sp	10±1	10±1	7±1	17±1	12±1	7±1	18±1	16±1	12±1
P. aeruginosa	18±1	15±1	12±1	10±1	6±1	5±1	15±1	9±1	8±1
B.cereus	24±1	21±1	18±1	10±1	12±1	14±1	9±1	8±1	6±1
Trichosporon spp	47±1	45±1	42±1	28±1	18±1	14±1	40±1	35±1	30±1

strains	Diameter of inhibition zones (mm) [a]						
	Carum carvi L			Syzygium aromaticum			DMSO" control
	Dilutions			Dilutions			
	1/2	1/4	1/8	1/2	1/4	1/8	
S.aureus	20±1	18±1	12±1	29±1	26±1	25±1	0
E. coli	7±1	7±1	4±1	14±1	12±1	11±1	0
Salmonella thyphimurium	15±1	13±1	11±1	16±1	12±1	13±1	0
Shigella sp	10±1	10±1	9±1	13±1	11±1	10±1	0
P. aeruginosa	15±1	16±1	17±1	6±1	6±1	3±1	0
B.cereus	9±1	6±1	4±1	16±1	14±1	17±1	0
Trichosporon spp	34±1	33±1	28±1	29±1	25±1	24±1	6

[a]: well diameter included

Based on the values in Table 3, we have classified the pathogenic strains according to their sensitivities to essential oils.

Table 4: Grading of stacks according to their sensitivity to HE

Strains	*Thymus vulgaris*	*Salvia officinalis*	*Allium sativum*	*Myrtus communis L*	*Rosmarinus officinalis*	*piper nigrum*	*Carum carvi L*	*Syzygium aromaticum*
S.aureus	+++	+++	++	+++	++	++	++	+++
E. coli	+++	++	+	+	+	+	-	+
Salmonella thyphimurium	+++	+	++	+	+	++	+	+
Shigella sonnei	+++	+++	++	+	++	++	+	+
P. aeruginosa	-	+	+	++	+	+	++	-
B.cereus	+++	++	+	+	-	-	+	++
Trichosporon spp	+++	+++	+++	+++	++	+++	+++	+++

+++: D>20mm: extremely sensitive ; + : 9<D<14 : sensitive

++: 15<D<19: very sensitive   - D<8: resistant

⚜ *Thyme (Thymus vulgaris) essential oil against pathogenic strains*

1-factor ANOVA

Strains

	Sum of squares	ddl	Average square	F	Meaning
Intergroup	69.000	10	6.900	4.600	.012
Intra-group	15.000	10	1.500		
Total	84.000	20			

Estimated distribution parameters

	Strains	Dilutions	inhibition diameter
Minimum	1	1.000	7.00
Even distribution Maximum	7	7.000	48.00

Observations are not weighted.

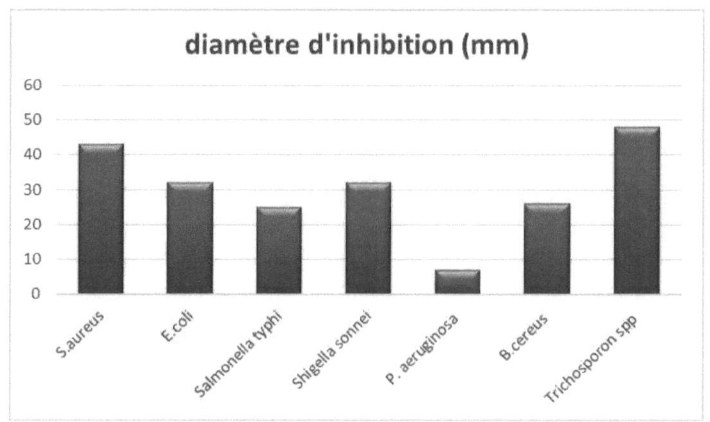

Figure 10: Histogram comparing inhibition zones for thyme EO

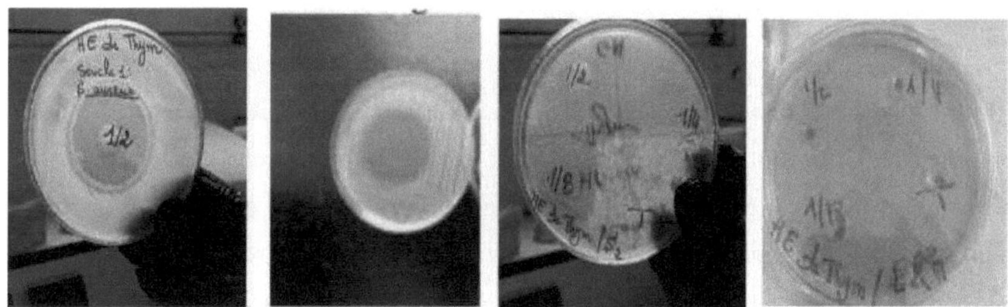

Figure 11: Thyme EO aromatogram

According to the above results

$P < 0.05$ so there was a significant difference in the inhibitory power of thyme essential oil against the 7 strains. This is explained by the fact that thyme EO had strong antibacterial activity against all pathogenic bacteria, with inhibition zones ranging from 10 to 48 mm.

In fact, the bacteria most sensitive to thyme EO were *S. aureus*, followed by *E. coli, Shigella sonnei, B. cereus and S. typhi*. While Ps. aeruginosa showed resistance to thyme EO as it has a higher level of Mg2+ in its outer membrane. A higher level of magnesium in the membrane will increase cross-linking between LPS and therefore reduce porin size and limit the migration of antimicrobial molecules across the bacterial membrane.

Strains of the genus *Pseudomonas* proved more resistant to HE. This resistance is not surprising. This has been confirmed by several previous studies (Hammer et al, 1999 Dorman and Deans, 2000).

Staphylococcus aureus, Escherichia coli, S.typhi, Shigella sonnei show a decrease in inhibition diameter parallel to the decrease in essential oil concentrations. Pseudomonas aeruginosa showed a constant inhibition diameter.

The fungal strain Trichosporon spp is extremely sensitive to thyme EO with a diameter of around 48mm, proving that thyme oil has fungicidal power even after dilution (1/2, 1/4 and 1/8).

Our results are proven by the literature studies have indicated that thymol and carvacrol possess an antimicrobial effect against a broad spectrum of bacteria: *Escherichia coli, Bacillus creus, Listeria monocytogenes, Salmonella enterica, Clostridium jejuni, Lactobacillus sake, Staphylococus aureus and Helicobacter pyroli* (rahmouni, 2014) [28]

(Abbas et al, 2016)[29] showed that *T. vulgaris* EO exhibited remarkable antibacterial activity for different bacterial strains except for *P.aeruginosa* with a diameter of 9mm which approaches that of ciprofloxacin. They found that thymol and carvacrol were the main components of this oil to exhibit this antimicrobial activity. The same authors demonstrated that these phenols increase bacterial cell membrane permeability and reduce protomotor force, thus lowering the intracellular level of ATP, which provides the energy required for the cell's chemical and metabolic reactions.

Thymus vulgaris essential oil has been shown to inhibit the growth o f a number of fungal strains, including *Candida albicans, Cryptococcus neoformans, Aspergillus, Saprolegnia and Zygorhynchus*. The same oil could potentiate the antifungal effect of amphotericin B against *C. albicans* **(Giordani et al. 2004; Pina-Vaz et al. 2004)**[30][31]

⚜ *Sage (Salvia officinalis) essential oil against pathogenic strains*

1-factor ANOVA

Strains

	Sum of squares	ddl	Average square	F	Meaning
Intergroup	54.583	13	4.199	.999	.527
Intra-group	29.417	7	4.202		
Total	84.000	20			

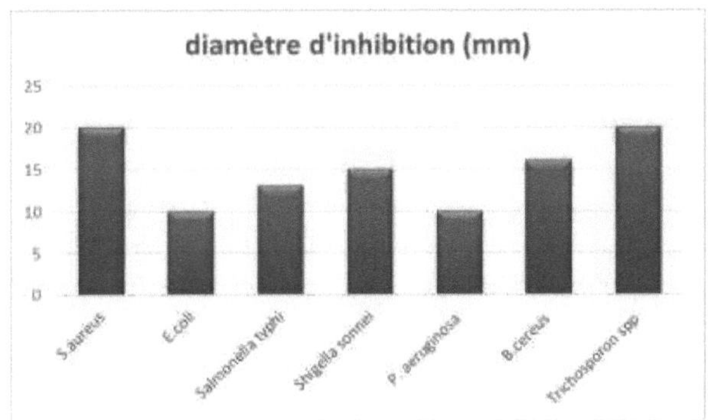

Figure 12: Comparative histogram of sage EO inhibition zones

p>0.05, which means that sage EO is slightly to moderately inhibitory to pathogenic strains, since the diameter of the inhibition halo is between 10 and 20mm.

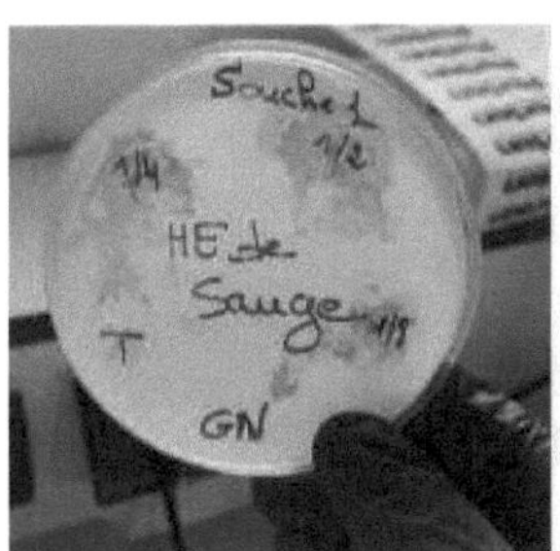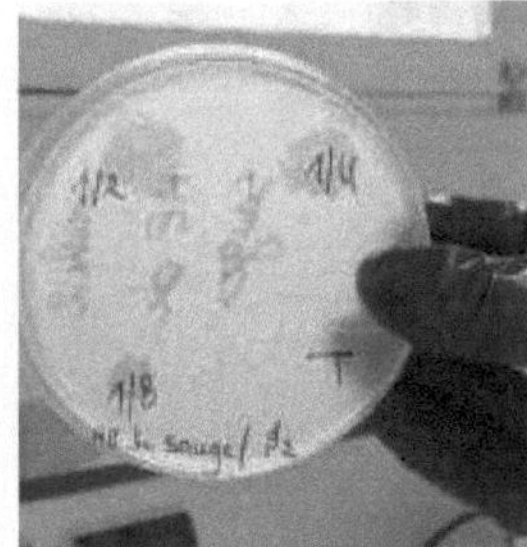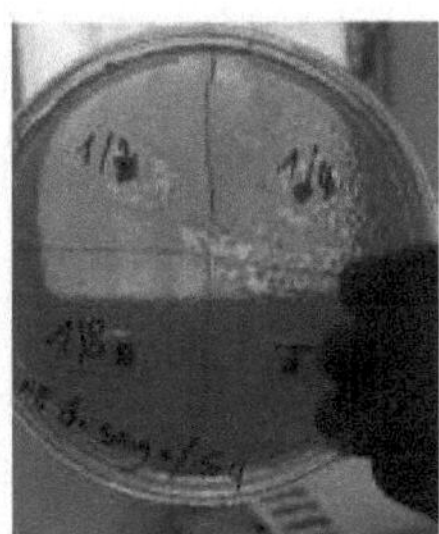

Figure 13: Sage EO aromatogram

The results of the aromatogram test show a variation in the inhibition efficacy of *Salvia Officinalis* essential oil against the bacterial strains tested: *Pseudomonas aeruginosa, Escherichia coli and Staphylococcus aureus, E. coli …*

Sage essential oil exerted weak inhibition against *Escherichia coli* and *P.aeruginosa, with a* 10 mm diameter zone of inhibition. In contrast, the essential oil of *S. officinalis* inhibits the growth of S.aureus and the yeast Trichosporon spp, with a 20 mm diameter zone of inhibition.

According to the classification of (Ponce et al.2003) [32], the zones of inhibition, varying between 8 and 20 mm, indicate that the strains tested are more or less sensitive to the essential oil of *Salvia officinalis* leaves. EOs have an effect on bacterial growth, preventing multiplication and toxin synthesis.

⚜ *Garlic (Allium sativum) essential oil against pathogenic strains*

1-factor ANOVA

Strains

	Sum of squares	ddl	Average square	F	Meaning
Intergroup	65.250	9	7.250	4.253	.014
Intra-group	18.750	11	1.705		
Total	84.000	20			

Variance homogeneity test

Strains

Levene statistics	ddl1	ddl2	Meaning
4.278a	6	11	.018

a. Groups with a single observation are ignored when calculating the homogeneity of variance test for strains.

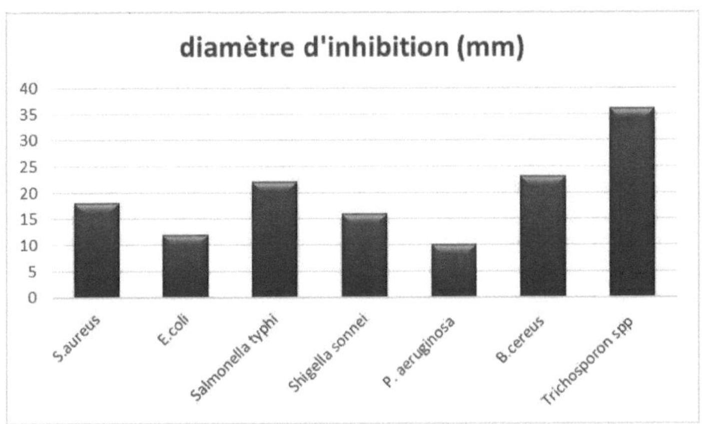

Figure 14: Histogram comparing inhibition zones for garlic EO

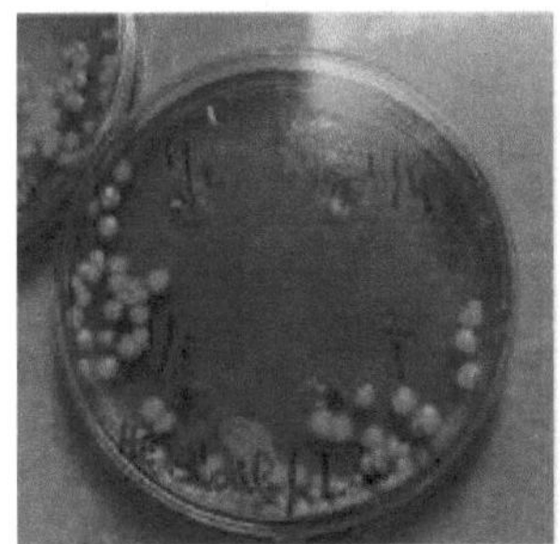
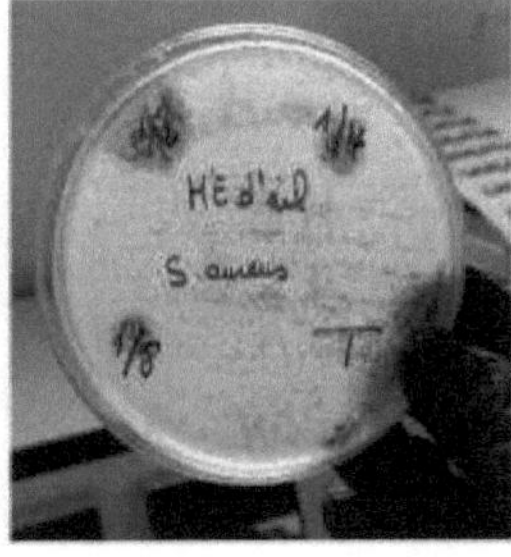

Figure 15: Garlic EO aromatogram

$P<0.05$ therefore there is a significant difference in the sensitivity of the strains to garlic oil, with diameters ranging from 10 to 36 mm, given that garlic oil has an extremely inhibitory effect on the fungal strain, hence the use of garlic oil (D=36).

⁜ *Myrtle (Myrtus communis L) essential oil against pathogenic strains*

1-factor ANOVA

Strains

	Sum of squares	ddl	Average square	F	Meaning
Intergroup	51.667	1	4.697	1.307	.349
Intra-group	32.333	9	3.593		
Total	84.000	20			

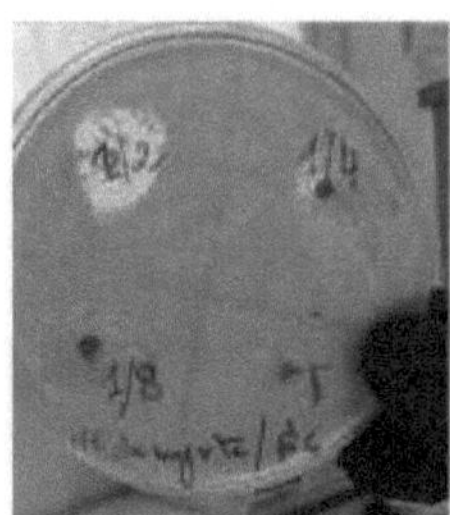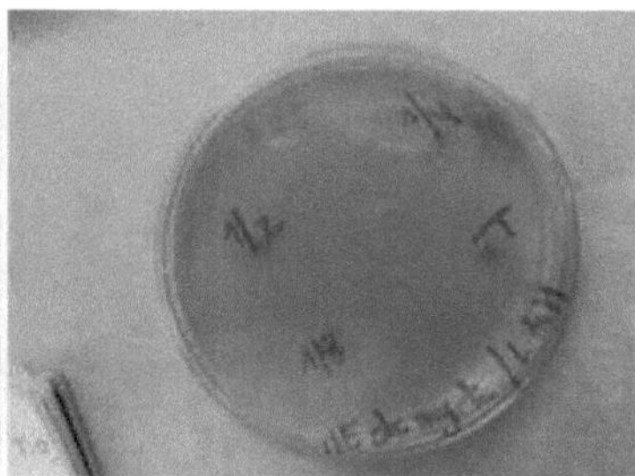

Figure 16: Myrtle EO aromatogram

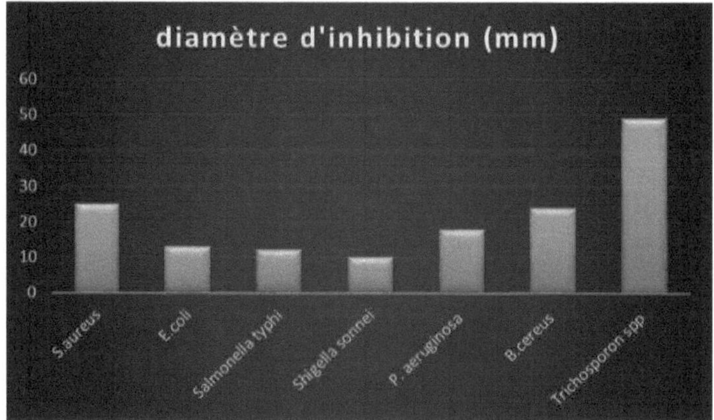

Figure 17: Histogram comparing inhibition zones for myrtle EO

p>0.05 there was no significant difference in myrtle oil's inhibitory activity on all pathogenic strains, i.e. myrtle oil inhibited the growth of Gram+ and Gram-.

Analysis of the results for the essential oil of common myrtle shows very high inhibitory activity against *Trichosporon spp*, with a relatively large zone of inhibition between (39-49mm).as the concentration of essential oil increases, so does the diameter of the ZI.

These results concur with those found by Touabia (2011)[35] and the work of Chebaibi et al. (2016)[33] on Moroccan common myrtle essential oil, showing that common myrtle EO has significant anti-candidosis properties. Suppakul et al. (2003)[36], have suggested that the antifungal activity of EOs can occur via two different mechanisms

Some constituents cause electrolyte leakage and depletion of amino acids and sugars, while others can be inserted into membrane lipids, resulting in loss of membrane functions.

According to Cox et al. (2000)[34], t h e antifungal action o f essential oils against *Candida albicans* is due to an increase in plasma membrane permeability, followed by rupture of the membrane, resulting in leakage o f cytoplasmic contents and consequent yeast death.

✦ Rosemary (Rosmarinus officinalis) essential oil against pathogenic strains

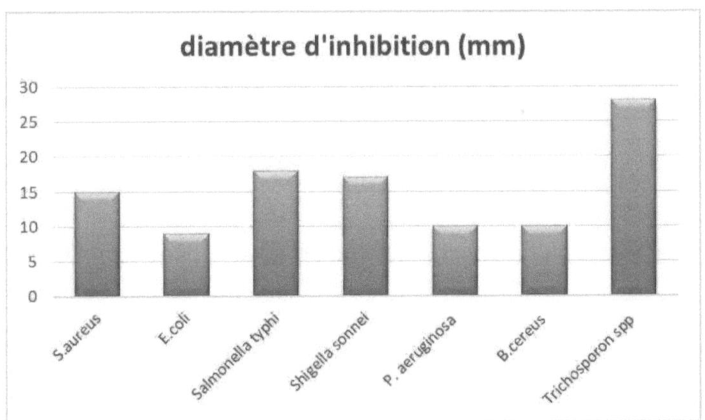

Figure 18: Histogram of rosemary EO

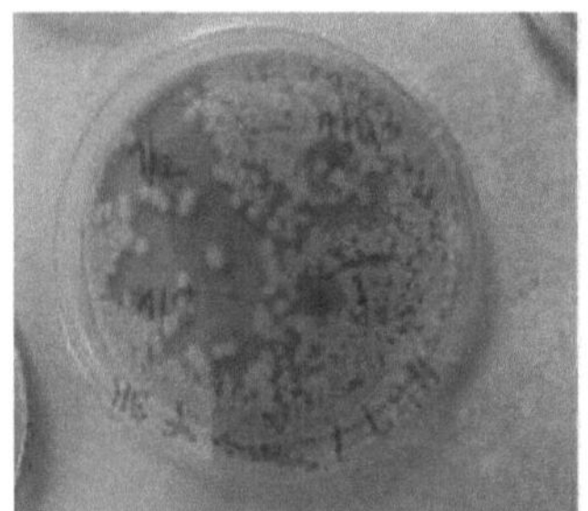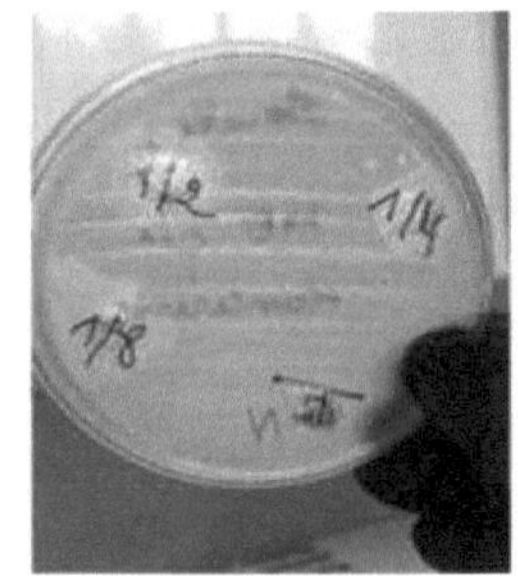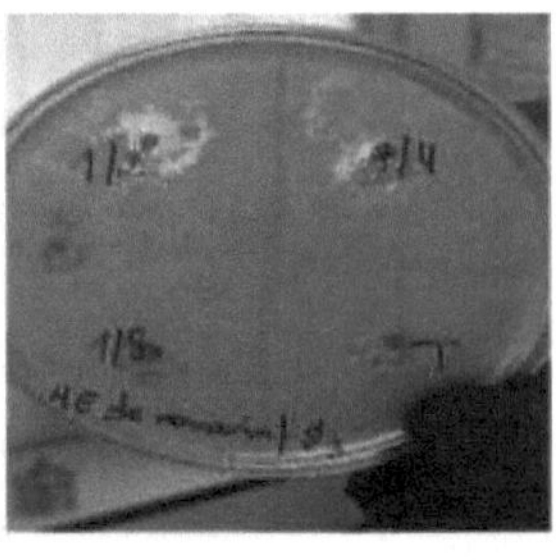

Figure 19: aromatogram of rosemary EO

According to the histogram, the diameters of the rosemary EO inhibition zones vary between 10 and 15mm for Gram+ and Gram- bacteria.

Despite the existence of weak, average zones of inhibition, our results prove the existence of antimicrobial activity against all seven strains tested, and the inhibitory effects increase considerably with EO concentration.

According to the classification of Ponce et al (2003), the zones of inhibition, varying between 15 and 19mm, indicate that the *Baccilus sp* strain is very sensitive and that the other strains tested are sensitive to rosemary essential oil (zone of inhibition between 8 and 14mm).

↓ Black pepper (piper nigrum) essential oil against pathogenic strains

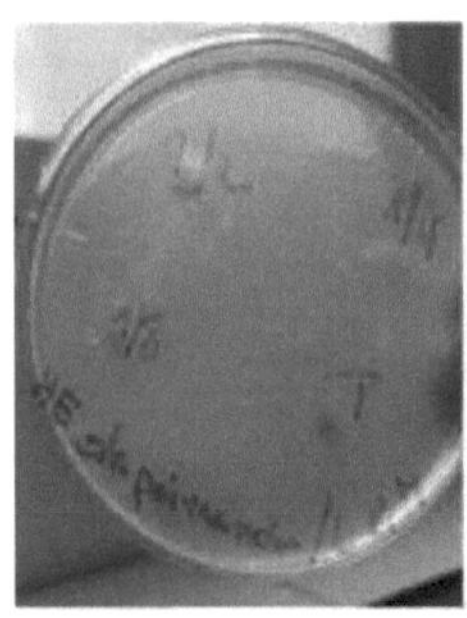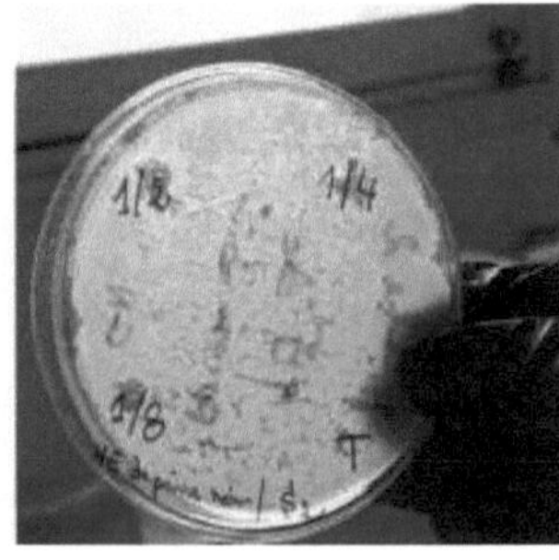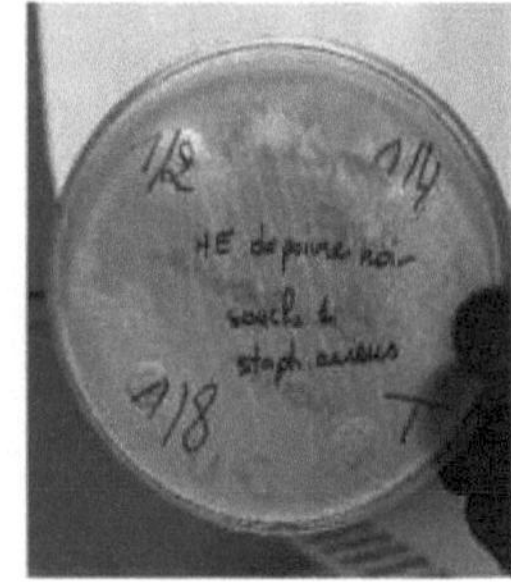

Figure 20: aromatogram of black pepper EO

Following these results, Vani et al (2009) showed that black pepper extracts have good antibacterial activity, indicating the presence of alkaloids, volatile oil, mono- and polysaccharides and resins. Alkaloids such as piperine, volatile oil and resins may be responsible for the antibacterial activity.

Studies have shown that black pepper's mechanism of action on bacteria (antibacterial) by altering cell membrane permeability due to the leakage of intracellular materials can lead to cell death.

Shiva et al (2013) reported antibacterial activity against all bacteria tested, with a zone of inhibition ranging from 8 mm to 18 mm. The zone of maximum inhibition was against Gram-positive bacteria *S. aureus* (18 mm) and *B. subtilis* (14 mm), while the bacteria Gram-negative *P. aeruginosa* (9 mm) and *E.coli* (8 mm). The maximum zone of inhibition was 100ul for all bacterial cultures. This indicates that the zone of inhibition increases with increasing black pepper concentration.

Shiva et al.2013, noted that the zone of maximum inhibition was against *S.aureus* bacteria (18mm) than *E.coli* bacteria (8mm).they found that *Piper Nigrum* extract, presented a greater activity against gram positive bacteria than gram negative bacteria.

From Vani et al (2009), they worked on the antibacterial activity of piper nigrum against certain Gram-positive (*S.aureus* ; *B.cereus* ; *S.faecalis*), and Gram-negative (*E. coli* ; *P.aeruginosa* ; *K.pneumoniae* ; *S. typhi*) pathogenic bacteria. they mentioned that :

Acetone extract of black pepper has excellent inhibition of the growth of Gram-positive bacteria, with **S.aureus** being the most sensitive, followed by B.cereus and Streptococcus. Among Gram-negative bacteria, **P. aeruginosa** was most sensitive to black pepper, followed by **E. coli, K.pneumoniae** and **Salmonella**.

Black pepper EO is highly antifungal against yeast.
Trichosporon spp. 40 mm in diameter.

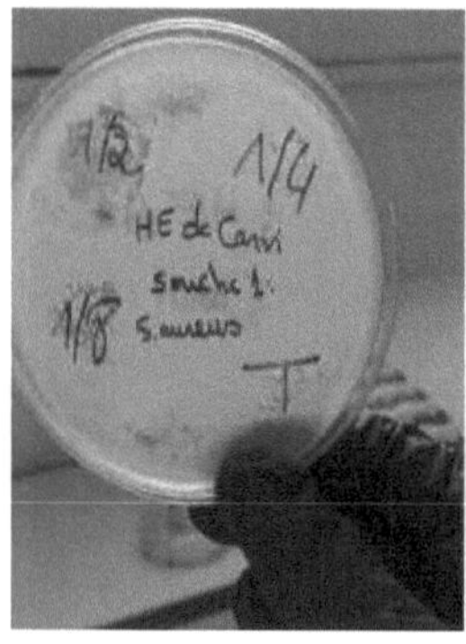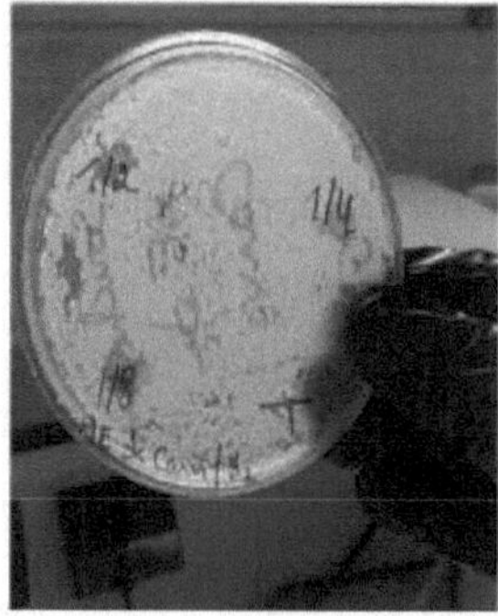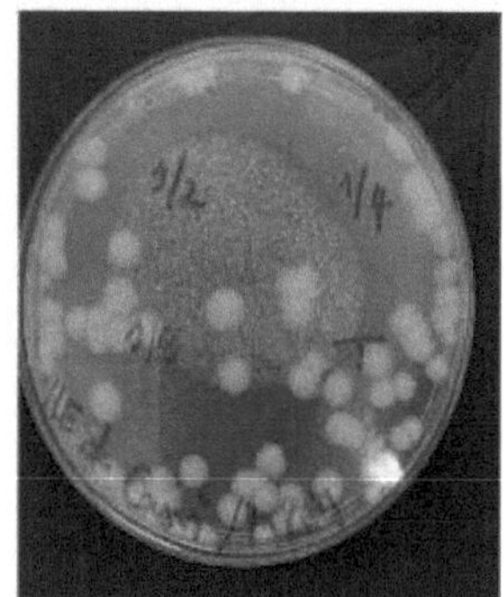

Figure 21: Caraway EO aromatogram

Carum carvi EO showed good antimicrobial and antifungal activity: the results show that the strains tested had the same sensitivity to this essence, this efficacy being due to the presence of the majority composition: limonene and carvone. Generally speaking, it has been reported that the presence of 1,8-cineole and carvone in H.E. are responsible for this antimicrobial activity.

The figure shows that Caraway oil is highly active against *Trichosporon spp.,* creating an interesting zone of inhibition around wells containing Caraway EO.

Based on the results obtained, we can say that the *Trichosporon spp* strain is characterized by a very high sensitivity to Carum carvi essential oil.

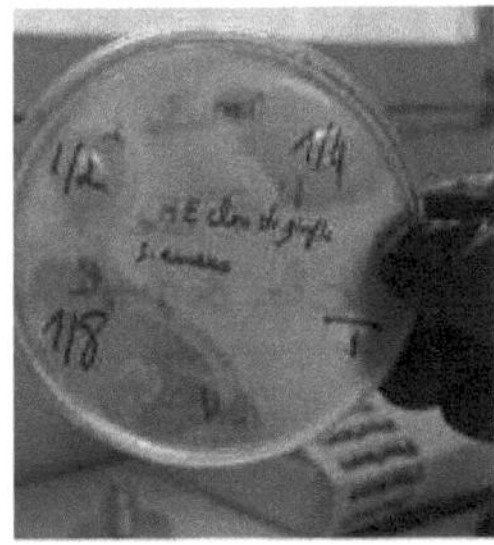
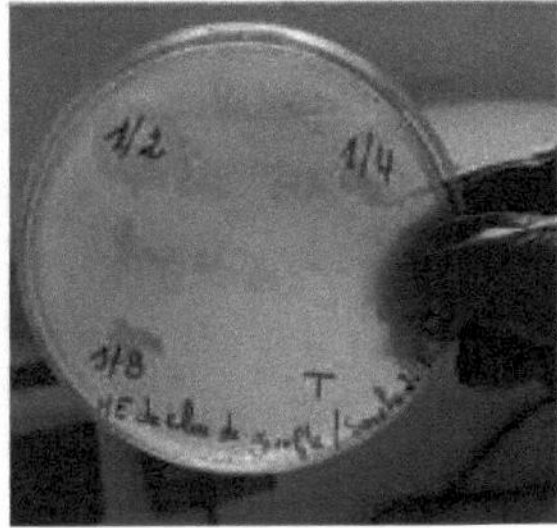
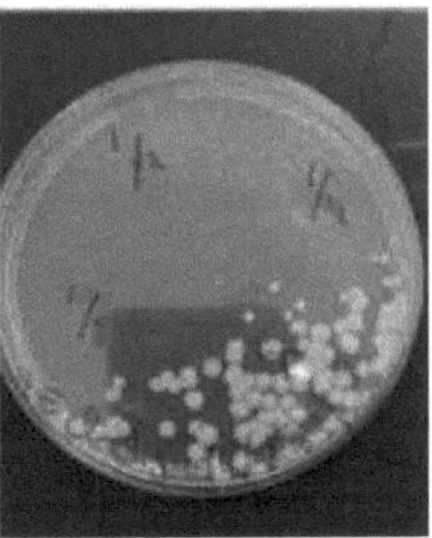

Figure 22: Clove EO aromatogram

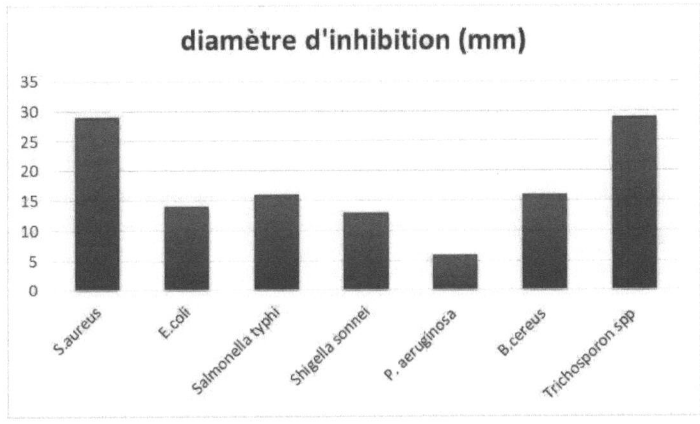

Figure 23: Clove EO histogram

Clove oil in graduated concentrations has a positive, i.e. sensitive, effect on bacterial strains: *E. coli and Staphylococcus aureus*; on the other hand, it is resistant to *Pseudomonas aeruginosa.*

Comparing the results for inhibition zone diameters, it's clear that gram-positive bacteria have larger inhibition zones than gram-negative bacteria for essential oil. *Staphylococcus aureus* had the largest inhibition diameter, while Pseudomonas showed no inhibition, making this strain highly resistant to our EO.

Essential oil concentration is related to zones of inhibition. The higher the concentration, the greater the zone of inhibition.

DMSO well results showed no antibacterial effect on bacterial strains due to the absence of inhibition zones.

Numerous studies have demonstrated that clove E.O. is highly antibacterial. This activity may be attributed to its main compound, "eugenol". Work by Valero and Giner in 2006 demonstrated that eugenol, among other compounds, inhibited bacterial growth. Furthermore, Rhayour's study (2016), showed that clove EO exerts its bactericidal activity mainly thanks to its majority constituent, eugenol, which belongs to the phenol family. It therefore seems that the bactericidal activity of EOs begins with the binding of these molecules to bacterial membranes, causing structural and permeability alterations that lead to the loss of cellular constituents and significant lysis of bacterial cells.

Regarding fungal strains, clove essential oil has good antifungal properties.

Clove essential oil has been shown to possess potent antifungal activity against opportunistic fungal pathogens, identical to the work of Eugénia and colleagues (2009) on *Candida albicans* and other fungal pathogens. Other studies have shown that clove EO, and Eugenol in particular, is highly fungicidal against *Candida albicans*. In the light of all this work and research, clove extract has a broad spectrum of antimicrobial activity, hence the importance of this oil as a highly effective preservative and antiseptic to prevent microbial development, especially when it comes to protecting health from the presence of pathogens.

1.3. Quantitative evaluation of the antibacterial activity of EO: MIC

The MIC is indicated by the well containing the lowest concentration of HE where there is 99% absence of bacterial growth.

In our study, we selected the strains most sensitive to essential oils and determined their MIC.

Table 5: Minimum inhibitory concentrations (MIC) of essential oils (%v/v) against selected strains

Essential oils	*S.aureus*	*E.coli*	*Salmonella thyphi*	*Trichosporon spp*
Thymus vulgaris	0.125	0.125	0.25	0.03
Salvia officinalis	0.06	0.25	0.25	0.06
Allium sativum	0.06	0.125	0.125	0.06
Syzgium aromaticum	0.125	1	0.5	0.06

Trichosporon spp is the most sensitive strain, with a cmi of 0.03 (%v/v) towards thyme essential oil

On the other hand, the *E. coli* strain showed moderate sensitivity to the essential oils tested, as did *Salmonella thyphi*.

Discussion

Orange juice is rich in water and sugars that are easily metabolized by microorganisms, making it a perfect growing medium (Forrest, 2001). These conditions easily expose these products to contamination by pathogenic fungi and bacteria. Single or multiple contamination of juice can come from many sources: soil, water, fertilizers (especially compost), workers, farm equipment and storage and processing conditions.

The flora present in juice is nothing other than the result of a variety of micro-organisms originating from the fruit itself, the storage atmosphere, sometimes the processing and packaging chain, water, air or even personnel.

Despite efforts to reduce the load of microorganisms in the industrial air or to attenuate the flora associated with fruit, the few bacterial or fungal cells that can escape control find in juice an environment entirely favorable to their growth and reproduction.

According to Cox et al (2000), the antifungal action of essential oils on Candida albicans is due to an increase in plasma membrane permeability, followed by rupture of the membrane, resulting in leakage of cytoplasmic contents and consequent yeast death.

The composition and antibacterial activity of the EOs were very different. A similar trend was observed by the authors, who showed that there was considerable variation between the antibacterial actions of essential oils. However, comparing the efficacy of oils between studies is difficult due to uncontrollable and external parameter differences. The composition of plant oils is known to vary according to climatic and environmental conditions. Moreover, antimicrobial properties can vary within the same plant.

Furthermore, some oils with the same common name may be derived from different plant species. Similarly, the method used to assess antimicrobial activity, and the choice of micro-organisms tested, differ from one publication to another. Agar and broth dilution methods are also commonly used.

These include differences in microbial growth, exposure time of micro-organisms to vegetable oil, oil solubility and the process used to solubilize or emulsify them. These and other factors may explain the large difference in MICs obtained by the agar dilution method in this study.

The prospects for this study include extracting essential oils and studying the antibacterial and antifungal properties of these oils on several microbial strains, with a view to possible disinfection of contaminated air. In addition to studying other biological properties of these plants, namely anti-inflammatory, antiviral, anti-lithiasic and other properties.

Conclusion

The aim of the present work is to find new natural products with antibacterial and antifungal properties against pathogenic strains that cause post-harvest citrus deterioration, and to compare the efficacy of these products with that of chemical fungicides used in this field.

Citrus fruits are rich in water and sugars that are easily metabolized by microorganisms. In the first part of this work, we isolated and identified the strains responsible for citrus deterioration; the results of their identification enabled us to recognize the following species: *P. aeruginosa, E. coli, Salmonella thyphi, S.aureus* and the fungal strain *Trichosporon spp*, all of which were found to cause citrus rot.

We studied the antifungal activity of thyme, sage, myrtle, clove, caraway, garlic and black pepper essential oils against isolated microorganisms. The results obtained show that all eight EOs give satisfactory results on the diametral growth of these strains, with a slight difference in efficacy with their volatile fractions such that thyme EO is the most effective given the presence of inhibition halos of diameter D>40mm in diameter against the fungal strain. This is explained by its high phenol content, 30 to 40% thymol and 5 to 15% carvacrol, which are responsible for its bactericidal, fungicidal and vermicidal activities, Sage EO is also the most effective in terms of inhibiting all the pathogenic germs identified in our study, while the other oils are less effective. Essential oils are therefore a good alternative to chemical pesticides, given their efficacy and benefits for human health, and even a 10-day follow-up showed them to be equally effective against pathogenic microorganisms.

References

(1)Benaissat F. ; 2015; la caractérisation de la sensibilité des variétés d'agrumes aux pourritures en post-récolte " .Université Sidi Mohammed Ben Abdellah , Faculté des Sciences et Techniques.

(2)Jacquemond C, Curk F, Zurru R, Ezzoubir D, Kabbage T, Luro F , et Ollitraut P , 2002-'Rootstocks: key components of sustainable citrus growing' , Session 3 .
Quality in the yard.

(3) DOUYLE M.P. &PADHYE V.V., 1989. Esherichia coli Food borne Bacterial pathogens.(eds.).Mareel Dekker Inc.

DOUYLE M.P. & CLIVER D.O., 1990. Esherichia coli, D.O. Cliver (ed.) Academic.Press, San Diago, California.

(4)FARBER, J.M., 1989. Food borne pathogenic microorganisms: Characteristics of the organisms and their associated diseases .I. Bacteria. Journal of the Canadian Institute of Food Science and Technology 22 (4): 311-321.

(5)M. Oussalah, S. Caillet, L. Saucier and M. Lacroix (2007). "Inhibitory effect of selectedhuile essential plants on the growth of four pathogenic bacteria: E. coli O157:H7, Salmonella Typhimurium, Staphylocoqueaureus and Lisaurait monocytogenes ," Food Control , vol. 18, pages 414 - 420

(6)Sabrine El Adab , Lobna Mejri , Imen Zaghbib , Mnasser Hassouna (2016) << Evaluation of Antibacterial Activity of Various Commercial Essential Oils >> American Journal of Scientific Research for Engineering, Technology and Science (ASRJETS) Volume 26 , No 3 , pp 212 -224

(7)   Imbert E., 2005. Mediterranean citrus. Fruitrop. Le point sur les agrumes méditerranéens.122 :6P

(8)Loussert R., 1987-Les agrumes. Techniques agricoles méditerranéennes. Paris: Tech. et Doc. Lavoisier.130P

(9)Aissa Dilmi Fadhila, Chaouchi Amira(2021). Microbiological study and the preservation of the agro-food quality of some varieties of citrus fruits grown in Algeria (Orange).

(10)   Redouane BASSAID OULHADJ (2020). Extraction of Lepidium sativum essential oil by several extraction methods [study of the effect of pretreatment on the yield and physicochemical characteristics of the oil].

(11)   HESSAS Thafsouth & SIMOUD Sounia (2018). Contribution to t h e study of the

chemical composition and evaluation of the antimicrobial activity of the essential oil of Thymus sp.(**2018**).

(12) Salah Benkherara, Ouahiba Bordjiba & Ali Boutlelis Djahra(2011). Study of the antibacterial activity of essential oils of Sage officinale: Salvia officinalis L. on some pathogenic enterobacteria.

(13) Secke C (2007). Contribution à l'étude de la qualité bactériologique des aliments vendées sur la voie publique de Dakar. Doctorat d'état en médecine vétérinaire. Cheikh AntaDiop University.

(14) FEDALA   Nazih.., MOKHTARI Moussa., MEKIMENE Lakhdar(2022).
contribution a la valorisation des dattes (deglet-nour) dans la vabrication du fromage de chevre. revue agrobiologia, 10,1918-1928.

(15) HAOUCHINE Lamia, KHENNACHE Ouardia.2017 Study of the antibacterial activity of extracts of four medicinal plants from Kabylie: Arbutus unedo L.,Phlomis bovei de Noé.,Rosa sempervirens L.and Verbascum sinuatum L.

(16) Khaled Attrassi, Bacteriological Quality of Citrus Fruits (Morocco)International Journal of Environment, Agriculture and Biotechnology, 5(2) Mar-Apr, 2020

(17) Alexandra MARTINS 2020. Antibacterial essential oils: the example of thyme (thymus)

(18) Peng. J , Tang J, Diane M., Shyam S., Anderson N., and Joseph R. (2015). Powers. Thermal pasteurization of ready-to-eat foods and vegetables: Critical factors for process design and effects on quality. Food Science and Nutrition, 57, 2970-2995.

(19) Deak T and Beuchat L, 1993 - Yeastsassociatedwith fruit juiceconcentrates. Journal of Food Protection, 56(9), 777-782

(20)   Nadjib Mohamed, FERHAT Amine, KAMELI Abdelkrim (2019). ESSENTIAL OIL EXTRACTION AND DISTILLATION METHODS. Review Agrobiologia, 9, 1653-1659.

(21)   AFNOR (Association Française de Normalisation). (1970). Determination of pH

(22) Lis-Balchin M., 2002, Lavender: the genus Lavandula, Taylor and Francis, London, p. 37, 40.

(23)   Abed Soumia, Messaadia Bouchra. Study of the physicochemical and biological properties of Thymus vulgaris L. 19/09/2021.

(24)   Fabian, D., Sabol, M., Domaracké, K., Bujnékovâ, D. (2006). Essential oils their antimicrobial activity against Escherichia coli and effect on intestinal cell viability. Toxicol. Invitro 20, 1435-1445.

(25) Paupardin C, Leddet C and Gautheret R. (1990). Genetics, selection and multiplication. Iamelioration of Artemisia species (Artemisia ubelliformis and E. genipi) by meristem culture. J. Jap. Bot. 65, 33.

(26) Djeddi S, Bouchenah N, Settar I- Composition and antimicrobial activity of essential oil of Rosmarinus officinalis from ALGERIA- Chemistry of Natural Compounds ; vol.43 :N)4.2007

(27) BOUALEM S, BOUMRAR Silia. Formulation of a disinfectant gel based on Rosemary (Rosmarinus officinalis L) essential oil and evaluation of its antimicrobial activity [Dissertation] Université Mouloud Mammeri Tizi Ouzou.2016

(28)   Rahmouni, M.(2014). Contribution à l'étude de l'activité biologique et la composition chimique des huiles essentielles de deux Apiacées (Ferula vesceritensis Coss et DR et Balanseagla berrima Desf.) Lange. Master's thesis, Université Ferhat Abbas - Sétif 1.Algérie.

(29) Abbas, N and Guerriche, F. (2016). Phytochemical study of thyme Thymus vulgaris.

L. (Lamiaceae) and insecticidal evaluation of its crude ethanolic extract against two insects, the harmful Aphis fabea and the beneficial Apis mellifera. Master's thesis,

Université M' hamed Bougara Boumerdès, Algeria.

(30) Giordani R., Regli P., Kaloustian J., Mikaïl C., Abou L., Portugal H (2004). Antifungal effect of various essential oils against Candida albicans. Potentiation of antifungal action of amphotericin B by essential oil from Thymus vulgaris. Phytother Res, 18(12), 990-995

(31) Pina-Vaz C., Gonçalves Rodrigues A., Pinto E., Costa-de-Oliveira S., Tavares C., Salgueiro L et al (2004).Antifungal activity of Thymus oils and their major compounds. J Eur Acad Dermatol Venereol, 18(1), 73-78.

(32) Ponce A.G., Fritz R., del Valle C. and Roura S.I., 2003, Antimicrobial activity of essential oils on the native microflora of organic Swiss chard, Lebensm.-Wiss.u.-Technol.36, p.679- 684.

(33) Chabaibi A., Marouf Z., Lahazi F ., Filali M., Fahim A., Ed-Dra .2016 Evaluation of the antimicrobial power of essential oils from seven medicinal plants harvested in Morocco.

(34) Cox S.D., Mann C.M., Markham J.L., Bell H.C., Gustafson J.E., Warmington T.R., Wyllie S.G., 2000.The mode of antimicrobial action of the essential oil of Melaleuca alternafolia (tea tree oil).Journal of Applied.Microbiology, Vol. 88, pp 170-175.

(35) Touaibia M., 2011. Contribution to the study of two medicinal plants: myrtus communis and myrtus nivellei Batt et Trab, obtained in situ and in vitro. Mémoire de Magister en Biologie. Université Blida .Algérie.175 p.

(36) Suppakul P., Miltz J., Sonneveld K. and Bigger S. W., 2003. Antimicrobial properties of Basil and its possible application in food packaging. J.Agric. Food Chem, Vol.51, pp: 3197- 3207.

APPENDICES

Appendix 1

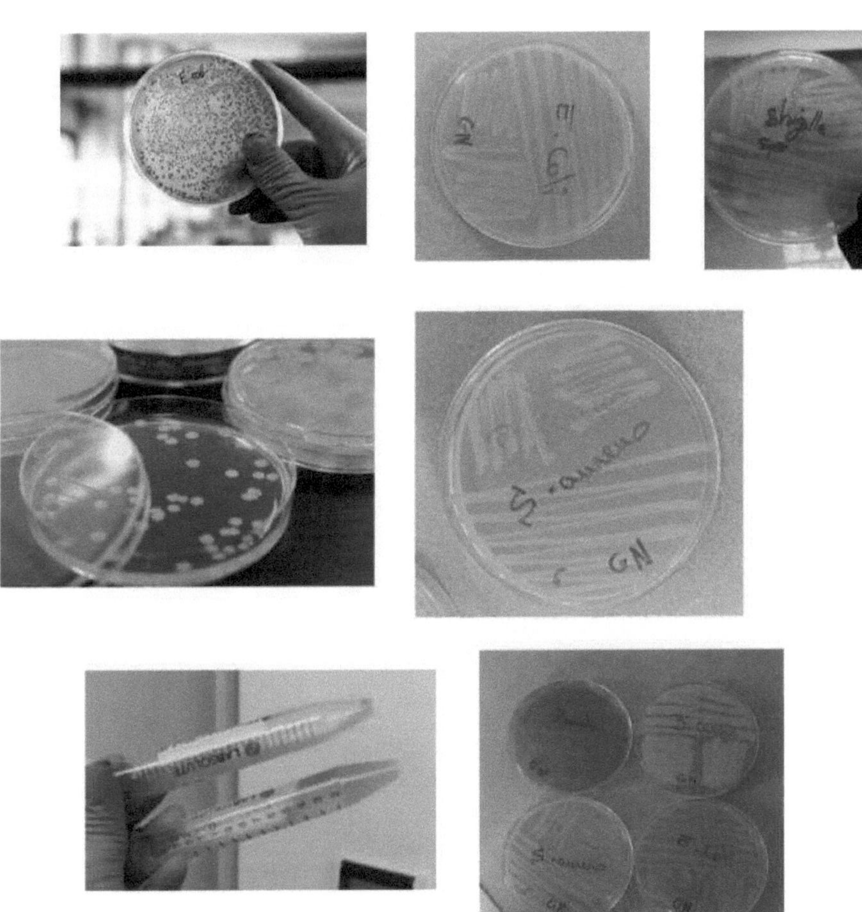

Coagulase test Re-isolation of strains

Appendix 2

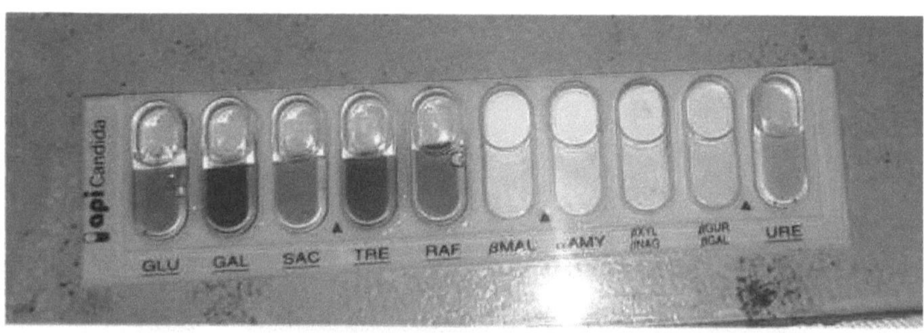

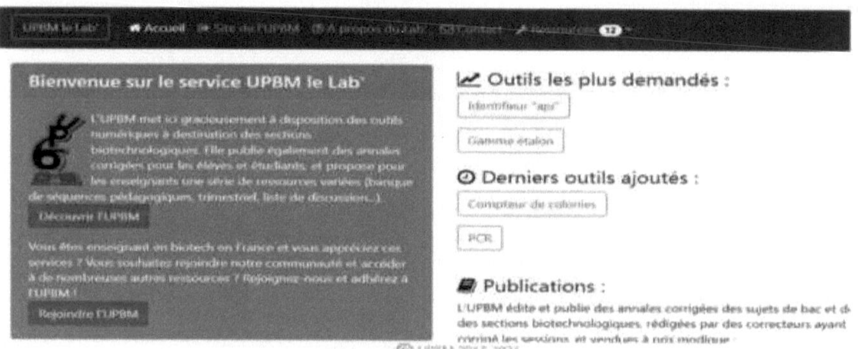

TABLEAU DE LECTURE

TESTS	COMPOSANTS ACTIFS	QTE (mg/cup.)	REACTIONS	RESULTATS	
				NEGATIF	POSITIF
1) GLU	D-glucose	1,4	acidification (GLUcose)		
2) GAL	D-galactose	1,4	acidification (GALactose)		
3) SAC	D-saccharose	1,4	acidification (SACcharose)	violet gris-violet	jaune vert / gris
4) TRE	D-trehalose	1,4	acidification (TREhalose)		
5) RAF	D-raffinose	1,4	acidification (RAFfinose)		
6) ßMAL	4-nitrophényl-ßD-maltopyranoside	0,08	ß-MALtosidase	incolore	jaune pâle-jaune vif
7) αAMY	2-chloro-4-nitrophényl-αD maltotrioside	0,168	α-AMYlase	incolore	jaune pâle-jaune vif
8) ßXYL	4-nitrophényl-ßD-xylopyranoside	0,095	ß-XYLosidase	incolore-jaune très pâle / bleu / vert **	jaune pâle-jaune vif
9) ßGUR	4-nitrophényl-ßD-glucuronide	0,063	ß-GlUcuRonidase	incolore / bleu / vert	jaune pâle-jaune vif
10) URE	urée	1,68	UREase	jaune-orange pâle	rouge
11) ßNAG (dans tube n° 8) *	5-bromo-4-chloro-3-indoxyl-N-acétyl-ßD-glucosaminide	0,09	N-Acétyl-ß-Glucosaminidase	incolore / jaune	bleu / vert **
12) ßGAL (dans tube n° 9) *	5-bromo-4-chloro-3-indolyl-ßD-galactopyranoside	0,0815	ß-GALactosidase	incolore / jaune	bleu / vert

* Les tubes 8 et 9 sont bifonctionnels :   tube 8 : ßXYL (test n° 8) / ßNAG (test n° 11)
tube 9 : ßGUR (test n° 9) / ßGAL (test n° 12)

** Toute trace verte dans la cupule 8 = ßXYL (–) ßNAG (+)

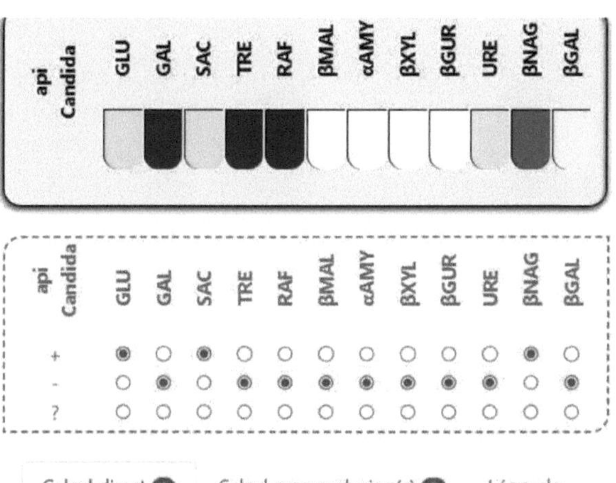

Calcul direct ①   Calcul avec exclusion(s) ⑧   Légende

Les calculs proposent (cliquez sur ⊕ pour voir les détails du profil) :

1. **Trichosporon spp 2** ⊕ avec une probabilité de 99.9 % (excellente identification)

99.9%

Les taxons ayant une probabilité trop faible (< 5%) sont éliminés

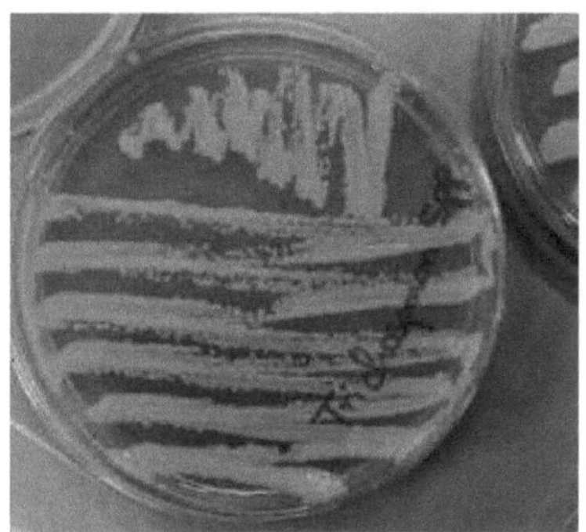

Trichosporon spp

I want morebooks!

Buy your books fast and straightforward online - at one of world's fastest growing online book stores! Environmentally sound due to Print-on-Demand technologies.

Buy your books online at
www.morebooks.shop

Kaufen Sie Ihre Bücher schnell und unkompliziert online – auf einer der am schnellsten wachsenden Buchhandelsplattformen weltweit! Dank Print-On-Demand umwelt- und ressourcenschonend produziert.

Bücher schneller online kaufen
www.morebooks.shop

Printed by Books on Demand GmbH, Norderstedt / Germany